METEOROLOGY FOR SEAFARERS

METEOROLOGY FOR SEAFARERS

BY

LIEUTENANT-COMMANDER R. M. FRAMPTON, R.N., F.N.I., F.R.Met.S.

AND

P. A. UTTRIDGE, B.Sc, M.Sc, F.R.Met.S.

GLASGOW

BROWN, SON & FERGUSON, LTD., NAUTICAL PUBLISHERS

426 DRUMOYNE ROAD

First Edition	–	1988
Second Edition	–	1997
Third Edition	–	2008
Reprinted	–	2010
Fourth Edition	–	2015
Reprinted	–	2016
Fifth Edition	–	2017

Front Cover:–Geostationary Satellite Visible Image, Meteosat Seviri–1500 UTC 21st August 2015.

ISBN 978-1-84927-074-8
ISBN 978-1-84927-056-4   (Fourth edition)

©2017–BROWN, SON & FERGUSON, LTD., GLASGOW, G51 4DA
Printed and Made in Great Britain

PREFACE

Commander C. R. Burgess, former Meteorological Officer in the UK Meteorological Office, Bracknell, and in the Royal Navy, completed Meteorology for Seamen in 1950 soon after he joined The Marine Society as its Secretary. His book was a standard work for some 30 years, combining the factual presentation of the subject with the then popular question and format. A further revision became necessary soon after his death in 1982. As many advances had been made in Meteorology and the presentation of textbooks so changed, it was decided to present a completely rewritten and revised text with the most up to date illustrations available, which culminated in the publication of the first edition of Meteorology for Seafarers, in 1988.

The text aims to present the fundamentals of Meteorology, and highlight those aspects which are of particular interest to all seafarers. In doing so, it does not aim to provide a simple explanation, as this is regularly and professionally done by the radio, television and more elementary textbooks, nor does it delve into the highly complex explanations presented by research papers. While the interaction of the seas and the atmosphere is considered, no attempt has been made to treat this aspect in detail, as there are many excellent works to which the seafarer can refer to improve his understanding.

Since Meteorology for Seafarers was first published, there have been four further editions which have successively kept the book up to date. In this edition, the fifth, opportunity has been taken to incorporate more recent examples of weather bulletins and charts, and selected figures have been modified. In addition further illustrative material has been included to support aspects noted in the text of Chapter 9 Tropical and Subtropical Circulation, and in the section in Chapter 11 on the Utilization of Facsimile Data.

Meteorology for Seafarers is therefore a technical book which aims to explain the complexities of the atmosphere, and to provide the information needed by professional seafarers studying for their First Class Certificates of Competency, and students for their degrees. If, at the same time, it encourages the reader to investigate and understand more clearly the forces of nature which affect his daily life, then Meteorology for Seafarers will have achieved the full ambitions of its authors.

ACKNOWLEDGEMENTS

The authors and publishers are indebted to and have great pleasure in acknowledging the help, guidance and hard work of so many in the preparation of this book and in particular:

Captain G. V. Mackie, Captain J. F. T. Houghton, Mr G. Allen, Captain S. Norwell, Ms S. North, and Mr D. Murphy of the UK Meteorological Office; Mr P. E. Baylis; Mr R. Barrett; Mr P. Bascombe; Captain E. H. Beetham; Mr P. Boemo; Mr N. Brown; Mr E. Cabrera; Mr J. Connaughton; Captain R. A. Cooper RFA; Captain J. C. Cox; Mr I. W. Cullen; Mr W. G. Davison; Mr W. T. L. Farwell; Mr N. D. Ferguson; Mr J. R. Gilburt; Professor P. Hardaker; Commander M. Hicks USCG; Captain F. Hugo; Dr F. A. James; Mr S. Johnson; Mrs E. Koo; Mr A. Lindberg; Mr C. R. Little; Mr L. McDermid; Captain D. M. McPhail; Captain J. McWhan; Captain S. D. Mayl; Mr C. D. Mercer; Professor J. Mitchell OBE FRS; Captain S. R. Montague; Mr M. Moore; Mr J. W. Nickerson; Mr R. K. Pilsbury; Mr D. G. Robbie; Ms A. Soares; Mr K. Takahata; Captain P. Thompson; Mr I. Thomson; Mr R. F. Williams; Mr D. Henderson; and Miss D. Durrant and Miss P. Musa who typed the script.

The following copyrights and sources are acknowledged with thanks:

Her Majesty's Stationery Office (HMSO) (Crown Copyright: UK Meteorological Office and UK Hydrographic Office): Figs. 2.1, 6.5, 7.9, 8.1, 8.3, 8.4, 8.6, 8.10, 8.14, 8.16, 9.3, 9.17, 9.20, 9.21, 10.2, 10.3, 10.9–10.12, 10.16, 10.18, 11.1, 11.2, 11.9–11.12. Tables 7.1, 7.2, 10.1, 11.3, 11.4, 11.5.

World Meteorological Organization: Figs. 10.4, 10.6, 10.7; Tables 6.1, 6.2, 10.2. Appendix 2–App. 2.1, 2.2.

US Mariners Weather Log and US National Weather Service, National Oceanic and Atmospheric Administration: Figs. 2.6, 9.9, 9.15, 10.14, 10.15, 10.17; Appendix 1–Table Al.

North American Ice Service US Coastguard International Ice Patrol: Fig. 11.5.

Bureau of Meteorology Australia: Figs. 8.11, 9.1, 9.19, 9.24, 10.13.

Japan Meteorological Agency: Figs. 9.2, 9.5, 9.6, 9.7, 11.6, 11.7, 11.8.

Swedish Meteorological and Hydrological Institute–Swedish Ice Service: Fig. 11.4.

The Nautical Institute–Seaways: Appendix 1.

NEODAAS/University of Dundee: Figs. 10.5(1), 10.5(2).

EUMETSAT, NEODAAS/University of Dundee: Figs. 9.14, 10.5(3), 10.5(4) and Front Cover.

Casella (London) Ltd: Fig. 4.7.

Furuno (UK) Ltd: Fig. 11.3.

Metworks Ltd: Figs. 10.20, 10.21.

The Science Museum: (London): Fig. 2.2.

R. G. Barry and R. J. Chorley: Atmosphere, Weather and Climate (Methuen): Fig. 4.5.

Mr J. Goulding: Fig. 11.13

Mr M. J. Leeson: Fig. 6.3.

J. G. Lockwood: World Climatology–An Environmental Approach (E. Arnold): Fig. 9.18.

Ms L. Mitchell and Mr A. Madge: Fig. 3.1

D. Riley and L. Spolton: World Weather and Climate (Cambridge University Press): Figs. 9.8, 9.22, 9.23.

Mr W. Wade Figs. 2.5(1)–2.5(3).

COLOUR PLATES

The authors received a very great number of colour photographs of cloud, sea states and meteorological phenomena, and are most grateful to the following contributors for permission to use their work (Plate №):

Captain S. J. Allen–12, 13, 14; Captain E. H. Beetham–21, 22, 23, 33, 34, 35; Mr P. Bascombe–39; Mr N. Brown–27, 42; Mr C. Doris–24; Mr J. R. Gilburt–19, 26, 36, 41, 43; Captain S. D. Mayl–25, 30; Mr M. Moore–10; Mr R. K. Pilsbury–2-2, 6; Captain C. R. Reed–16; Mr D. G. Robbie–1, 2-1, 11, 15, 31, 37, 38, 40; Captain J. F. Thomson–44.

CONTENTS

ILLUSTRATIONS AND PLATES

CHAPTER 10   ORGANIZATION AND OPERATION OF
METEOROLOGICAL SERVICES

COLOUR PLATES
METEOROLOGICAL PHENOMENA & SEA STATES

A – CLOUDS AND METEOROLOGICAL PHENOMENA

B – SEA STATES AND STORM WAVES

LIST OF TABLES

CHAPTER 1

THE ATMOSPHERE

INTRODUCTION

Meteorology is the scientific study of the atmosphere. The atmosphere is the envelope of gases surrounding the earth in which a number of processes varying in duration and dimension operate. These result in the *weather* experienced by an observer on the surface of the earth. *Climate* is the more general pattern of weather established by analysing, on an annual basis, the daily conditions at a particular point.

STRUCTURE AND COMPOSITION

The atmosphere may be considered as having a number of distinctive layers which are defined by the variation of air temperature with increase in height. For the standard atmosphere (Fig. 1.1) the layers are:

(a) *Troposphere* – Surface of the earth to 12 km – the *tropopause.*

(b) *Stratosphere* – Tropopause to 47 km – the *stratopause.*

(c) *Mesosphere* – Stratopause to 80 km – the *mesopause.*

Above the mesopause lies the *thermosphere,* a layer with a negligible quantity of gas, whose temperature increases with increase in height. It should also be noted that other layers may be defined on different criteria. The *ionosphere* for example, where the gases ionized by solar radiation affect the propagation of radio waves, exists from 60 km upwards.

Below the mesopause the mixture of gases in the atmosphere is nearly constant:

Gas		Percentage By Volume
Nitrogen	(N_2)	78.09
Oxygen	(O_2)	20.95
Argon	(Ar)	0.93
Carbon dioxide	(CO_2)	0.03

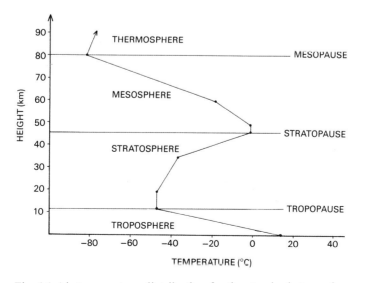

Fig. 1.1 Air temperature distribution for the standard atmosphere.

There are also traces of neon, helium, krypton, hydrogen and xenon. Ozone occurs at a greater level of concentration in the stratosphere, and this layer may be termed the *ozonosphere.*

In the troposphere, other chemical compounds are also present of which water vapour is by far the most significant. The amount of water vapour varies both in space and time, being greater in the lower troposphere and decreasing with increasing height. The volume of water vapour may be as much as 4% which, although small compared with the total volume of the atmosphere, is very significant in terms of the amount of energy which the atmosphere is able to store (Chapters 3 and 4). Other substances which may also be present are particles from the surface of the earth and from outer space, and chemical compounds which are manufactured within the atmosphere from the constituents present. Dramatic and sometimes unpleasant and damaging effects can be produced by these various substances; for example colourful sunsets and sunrises, red and acid rains.

Density and Pressure

The density of the gases decreases with increasing height (Fig. 1.2) since they are compressible, and 75% of the total mass of air is within the troposphere.

At the surface the *standard atmospheric pressure* is 1013.2 hPa (hectopascal) (Fig. 1.2), the equivalent of a tonne load on a man's shoulder. With increasing height atmospheric pressure decreases. This trend is partly related to the decrease of air density with increase in height. It should be noted that, while the value of the pressure at the surface in Fig. 1.2 relates to the *standard atmosphere,* actual values observed vary significantly from the standard (Chapter 2).

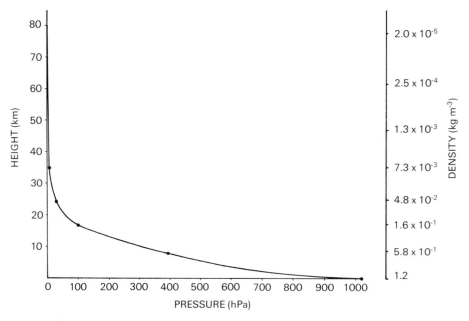

Fig. 1.2 Density and pressure distribution for the standard atmosphere.

Temperature

In the troposphere air temperature normally decreases as height increases. The slope of the line in Fig. 1.1 illustrates this change or lapse rate. As the slope relates to air temperature, it is termed the *Environmental Lapse Rate* (E.L.R.). For the standard atmosphere, where the air temperature at the surface is 15°C, the E.L.R. value in the troposphere is 6.5°C km^{-1}. The *tropopause,* the boundary of the troposphere with the stratosphere, is defined as the level at which the rate of decrease of temperature is 2°C km^{-1} or less, provided that the average decrease of temperature within the next 2 km does not exceed 2°C km^{-1}. In the standard atmosphere the tropopause is located at 12 km. However, in the actual atmosphere its height varies from 16 km at the equator to 9 km at the poles.

In the lower stratosphere, air temperature is constant with increasing height, but in the middle and upper parts it increases with increasing height. This temperature profile reflects the presence of ozone. In the mesosphere air temperature decreases rapidly with increasing height to −80°C at the mesopause.

CHAPTER 2

ATMOSPHERIC PRESSURE

INTRODUCTION

Atmospheric pressure is the most important meteorological element observed since it is the principal guide to the state of the atmosphere at a given time. The weather map or surface synoptic chart is derived from a set of readings of this element taken at different locations at internationally agreed times known as *synoptic hours*. The principal synoptic hours are 0000, 0600, 1200 and 1800 UTC.

DEFINITION

Pressure may be defined as the force which is exerted on unit area of a surface. The term *atmospheric pressure* therefore refers to the force which a column of air exerts on unit area of the earth's surface. The formula which defines this pressure is P=ρgh where P is the pressure, ρ the mean density of the air in the column, g the value of gravity, and h the height of the air column. The unit of measurement of atmospheric pressure is the *hectopascal* (hPa), this unit being equivalent to and having replaced the millibar (mb) (1 hPa = 1 mb = 10^2 Nm^{-2}, where N is a newton which is equal to 1 kg m^1 s^{-2}).

At any height above the surface of the earth the atmospheric pressure will be less than that at the surface. This is due to the smaller values of both the height of the air column and the mean density of air in the column (Fig. 1.2).

BAROMETERS

For many years the aneroid and mercury barometers have been used to determine atmospheric pressure. The precision aneroid barometer is considered to be the most accurate instrument for observing purposes.

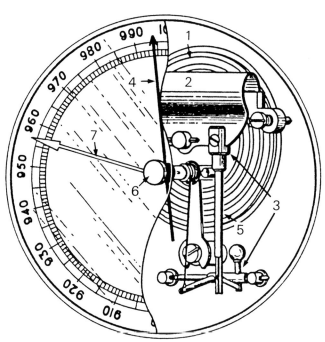

Fig. 2.1 Simple Aneroid Barometer.

Simple Aneroid Barometer

The principle on which all aneroid barometers are based is the use of the elastic properties of a metal to monitor changes in atmospheric pressure, and hence its value at a given time. With the simple aneroid barometer (Fig. 2.1) the metal, in the form of a corrugated capsule (1) almost exhausted of air, is compressed when the atmospheric pressure increases, and expands when the pressure decreases. The capsule is supported by a spring (2), and by means of a linkage system (3) its movement can be magnified. This is shown by a moving pointer (4) which rotates over a dial graduated in hectopascals. The metal from which the instrument is constructed is affected by changes in air temperature, which will result in an inaccurate reading. The effect is compensated for by leaving a small amount of air in the capsule and incorporating a bimetallic linkage (5). The elastic properties of the metal capsule change with time, and the instrument should therefore be checked at frequent intervals against an accurate barometer to determine its *index error*. This error is removed by means of the screw located in the back of the instrument.

The simple aneroid barometer can be used not only to record the atmospheric pressure to the nearest hectopascal at a given time, but also to observe the change in pressure over a period. By using the knurled knob (6) to set the pointer (7) at the current position of the moving pointer (4), the observer can note the change over the period by comparing the position of the moving pointer with the set pointer.

Precision Aneroid Barometer

The precision aneroid barometer (Fig. 2.2) is the instrument used to obtain readings of atmospheric pressure to the nearest tenth of a hectopascal, which are essential in the production of accurate synoptic charts. The accuracy of the reading is achieved by incorporating three aneroid capsules, and an improved magnification system and method of registering atmospheric pressure. The observer controls the magnification and registering of atmospheric pressure by using the external knob (A) to adjust the position of the micrometer screw (B). The position of the latter determines whether or not it is in contact with the contact arm (C). To obtain a reading the external knob is turned until contact between the micrometer screw and contact arm is made. The contact is indicated by a continuous line of light on the external display (D), when the button (E) is depressed. The external knob is then turned until the line of light just breaks, which indicates that the contact between the micrometer screw and the contact arm is just broken. At this point the readout (F) indicates the atmospheric pressure exerted on the capsules. Compensation for temperature variation of the instrument is provided by the inclusion of a small amount of air in the capsules. A pressure choke can be fitted (at position G) which restricts instrument response to short term changes of atmospheric pressure due to factors such as ship motion. The power supply for the electrical circuit is provided by a battery housed within the instrument, access being through the cover (H).

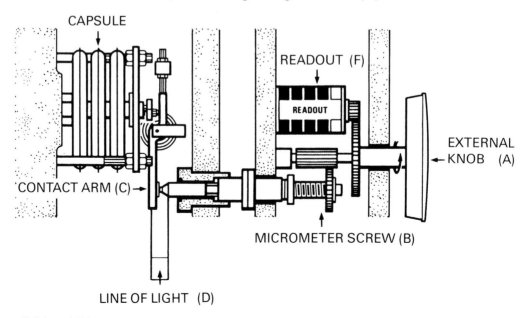

(1) Schematic Diagram.

Fig. 2.2 Precision Aneroid Barometer.

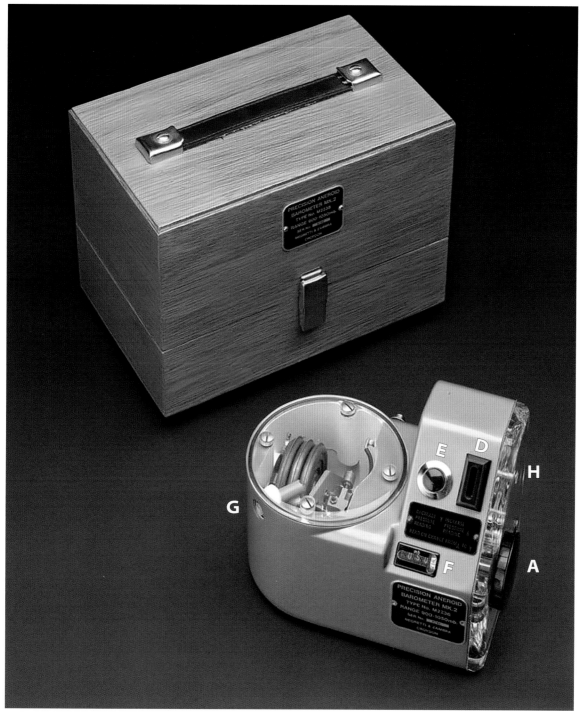

(2) General View.

Fig. 2.2 Precision Aneroid Barometer.

Corrections to Readings

All barometric readings which are to be used for synoptic purposes must be corrected to *standard datum* and for *index error*. The standard datum correction, often termed the height correction, is obtained from tables and requires the *height above mean sea level* of the instrument and air *temperature,* obtained from the marine screen (Chapter 3), to be known. The inclusion of air temperature takes into account the mean density of the air in the column between the height of the instrument and mean sea level. For a precision aneroid barometer the manufacturer supplies the index error correction which relates to errors inherent in the instrument. However, it is important that the instrument is checked frequently by comparing its readings with that of another barometer of known accuracy. Readings from a simple aneroid barometer are also corrected for height above mean sea level and index error if applicable.

ISOBARS

Corrected atmospheric pressure readings for a synoptic hour are plotted on a chart (Fig. 2.3), where they are compared with each other. Lines can then be drawn on the chart joining points having the same pressure value, each line being an *isobar* of a whole hectopascal value, e.g. 996 hPa. The interval between isobars on a chart is always constant, usually being 4 hPa (Fig. 2.3) or 5 hPa. The final isobaric chart thus shows the distribution of surface atmospheric pressure.

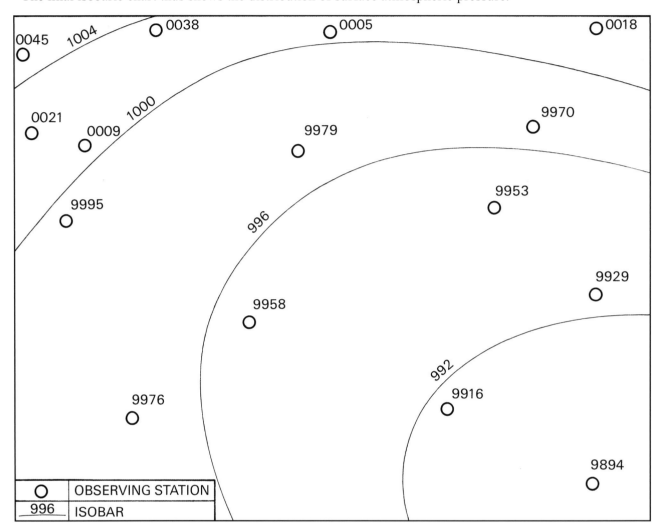

Fig. 2.3 Isobars.

The reading for each observing station is the corrected atmospheric pressure e.g. 9998 is 999.8 hPa.

PRESSURE TENDENCY

Atmospheric pressure may change over a period, and the change is termed the *pressure tendency.* For synoptic purposes the interval of time over which the change is observed is the 3 hours preceding the time of observation. The characteristics of the tendency observed may be described as "rising", "falling" or "steady", or a combination of these terms. The amount of change is also noted, being the difference between the reading at the beginning and end of the 3 hour period, to the nearest tenth of a hectopascal.

After plotting observed pressure tendencies on a chart, lines termed *isallobars* are drawn, which join points having the same pressure tendency. The value of an isallobar, either positive or negative, is to the nearest tenth of a hectopascal, and the interval between isallobars is constant. The isallobaric chart thus indicates areas where pressure changes have occurred, and may be used in developing the forecast.

BAROGRAPHS

The barograph has been designed to record atmospheric pressure and is used to determine the pressure tendency. This instrument works on the aneroid principle, having a number of capsules linked with each other and with a lever system connected to a pen arm. When the capsules respond to changes in atmospheric pressure, the pen arm moves either up or down across a scaled chart attached to a drum rotated by a clockwork mechanism. This record of pressure over a period is termed a *barogram.*

Fig. 2.4 Marine Barograph.

In the *Marine Barograph* (Fig. 2.4), the aneroid capsules are enclosed within a brass cylinder containing a silicone fluid. This arrangement acts as a "damping" mechanism, eliminating the response of the instrument to short term changes in pressure due to factors such as ship motion and vibration. However, the instrument is still free to respond to the long term changes of pressure. The clock operates for a period of 7⅓ days, and the drum is fitted with a 7 day chart which is changed once a week when the clock is wound.

DIURNAL VARIATION AND RANGE

When comparing a number of barograms of mid-latitude observations (Fig. 2.5(1) to (3)), it is particularly noticeable that there is neither a regular pattern nor a mean pressure value evident. However, in low latitudes a more regular pattern, termed the *diurnal variation* or daily change of atmospheric pressure, is usually evident (Fig. 2.5(4)). During a 24 hour period there are two maximum (at 1000 and 2200 LMT) and two minimum pressure values (at 0400 and 1600 LMT). In Fig. 2.5(4), the maximum pressure values occur at 0500 and 1700 UTC, the barograph on the vessel having been set on UTC at 91°E.

The *diurnal range,* the difference between the maximum and minimum readings, is usually small but noticeable in low latitudes, and in Fig. 2.5(4) the range lies between 3.5 and 4.0 hPa. As latitude increases, the diurnal range observed is smaller, with the result that in mid-latitudes the few tenths of a millibar are masked by the greater changes of pressure associated with pressure systems affecting the area (Fig. 2.5(1) to (3)).

In low latitudes the diurnal variation may be masked by excessive pressure changes due to tropical cyclones (Chapter 9), the decrease in pressure being far greater than the relatively small change attributed to diurnal variation (Fig. 2.6). Pressure tendency is therefore worth recording when in the tropics, for a change may well be the first sign of the approach of a tropical cyclone.

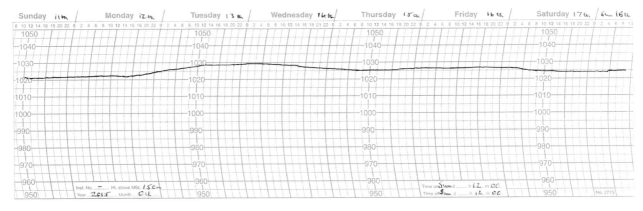

(1)

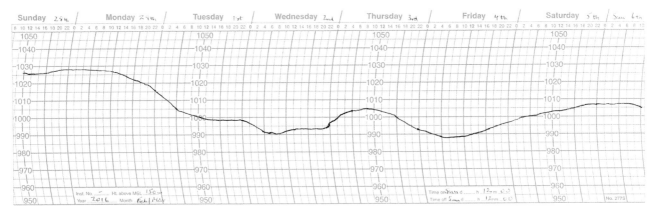

(2)

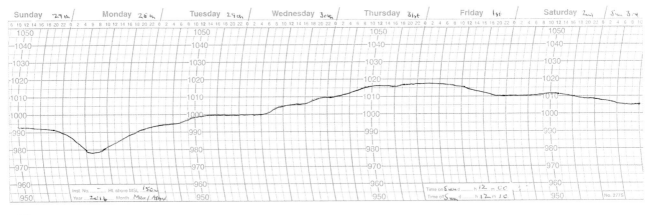

(3)

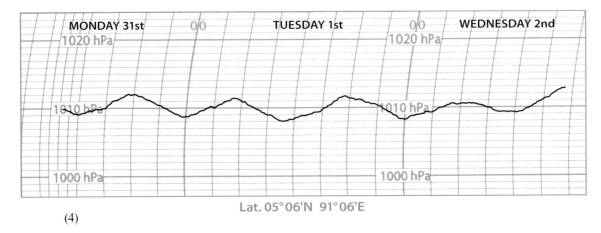

(4)

Fig. 2.5 Barograms: (1)–(3) Mid-latitude; (4) Low-latitude.

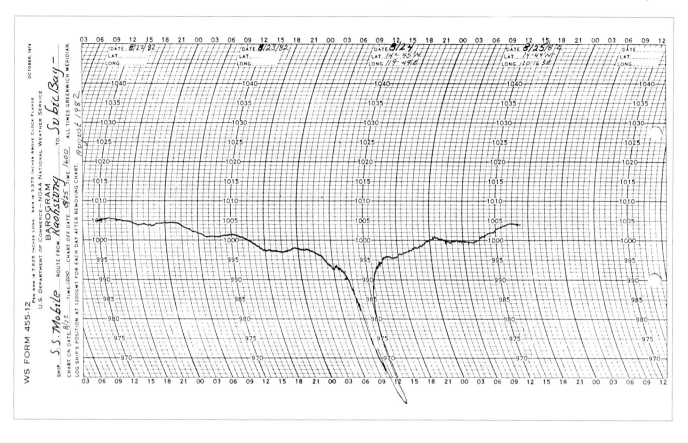

Fig. 2.6 Barogram – tropical cyclone (Typhoon Faye).

CHAPTER 3

TEMPERATURE

INTRODUCTION

Air temperature may be monitored at various heights above the surface. Surface air temperature is monitored at 1.25 m, the recommended height for instruments on land. Surface temperature is that monitored at the surface, and at land observing stations the surface may be turf, concrete or soil. At sea the surface temperature is monitored by voluntary observing ships or data buoys.

The air temperature recorded can in some circumstances reflect the influence of the underlying surface on a mass of air, and in particular the contrast which exists between land and sea surfaces. On other occasions the air temperature reflects the horizontal movement of air associated with pressure systems (Chapter 8).

OBSERVATION

Air Temperature

The standard instrument for monitoring air temperature is known as the *dry-bulb thermometer* which is a mercury thermometer enclosed in a glass sheath. The thermometer is housed in a *Stevenson Screen* which is designed to allow a free flow of air past the instrument, and at the same time eliminates the effects of solar and terrestrial radiation. Temperature is normally recorded to the nearest tenth of a degree Celsius (°C). A platinum resistance thermometer with a digital readout may be used as an alternative to a mercury thermometer.

The screen used at sea is the *Marine Screen* (Fig. 3.1), which is the Stevenson Screen modified to house the *Mason's Hygrometer* under seagoing conditions (Chapter 4). The screen is constructed of wood with louvred sides, one of which is a hinged door. The floor is slotted and the roof is a double structure, the inner one having ventilation holes. The entire structure is painted white both internally and externally.

Advances in technology have enabled alternative materials to be used in the manufacture of marine screens, such as a combination of durable ultra-violet stable plastic and treated aluminium. These materials enhance the life and reduce the amount of time spent on maintenance of the screen. The size and shape of screens have also changed with the expansion of automated sensors.

At sea 2 screens are normally used, one being on each bridge wing at a height of about 1.5 m above the deck and sited as far as possible from any of the vessel's heat sources. Observations of air temperature are always obtained from the screen located to windward.

Sea Temperature

Sea surface temperature is generally more difficult to monitor than either land surface or air temperature, and one or more methods may be used depending on the operation of the observing ship. If its deck height and speed permits, a sample of water is collected in a specially designed bucket made of canvas or rubber reinforced by canvas. The bucket is swung clear of the vessel to prevent any contamination of the sample, and submerged to a depth of about 1 m to avoid sea spray, which may be at a different temperature compared with that of the sea surface. Once inboard, the bucket is placed in a suitable position so that any energy transfer which may affect the temperature of the sample is minimized. A mercury thermometer provided with the bucket is used to record the temperature of the sample.

Fig. 3.1 Marine Screen.

If the above method cannot be used for operational reasons the temperature of the engine room intake is recorded. However, this data is liable to be inaccurate, partly due to the depth of the intake below the surface, and partly through the heating of the water sample as soon as it is inside the vessel.

Distant reading thermometers have been introduced for voluntary observing purposes. An electrical resistance unit is sited on the inside of the hull about 1 m below the waterline, where the hull temperature is similar to that of the sea. The sensing unit is connected to a digital readout located on the bridge, from which the observer can obtain a reading whenever required. This method minimizes the sources of error inherent in the other two techniques, but assumes the draught of the vessel will be more or less constant.

The temperature of the air or surface indicates its internal energy content. The processes involved in the gain or loss of this energy by the surface of the earth and the atmosphere are considered in the following sections.

SOLAR AND TERRESTRIAL RADIATION

The major source of energy driving the atmospheric circulation is the sun. The energy emitted by the sun is radiant energy or radiation in the form of electromagnetic waves, an important characteristic of which are their wavelengths. The sun, with a mean surface temperature of 6000 K, radiates energy within the range of wavelengths of 0.2 to 4 μm, with the maximum emission of energy being at the 0.5 μm wavelength (1 μm = 1 micrometre = 10^{-6}m). Thus *solar radiation* spans the ultraviolet, visible and infra-red parts of the spectrum.

The surface of the earth, with a mean temperature of 288 K (15°C), radiates energy in the infra-red part of the spectrum within the range of 4 to 100 μm wavelength. The maximum emission is at 10 μm wavelength. On comparing the wavelengths of radiation of the sun and the surface of the earth the terms *shortwave radiation* and *longwave radiation* may be applied respectively (Fig. 3.2).

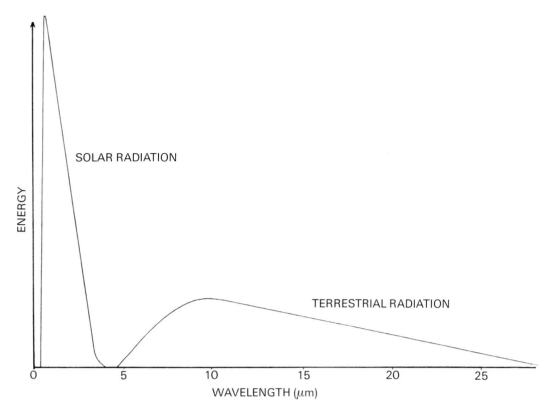

Fig. 3.2 Solar and terrestrial radiation.

The atmosphere also radiates energy. For the standard troposphere, within the range of temperatures noted (Fig. 1.1), the radiation emitted will have a similar range of wavelengths to that of the surface and is therefore longwave radiation. Together the radiation from these two sources is termed *terrestrial radiation.*

Solar Radiation and the Atmosphere

Five major factors affect the amount of solar radiation received at the surface of the earth:

1. Output of energy of the sun. This varies particularly in the amount of ultraviolet radiation which increases during a sunspot maximum.

2. Distance of the earth from the sun. At perihelion (minimum distance) the amount of solar radiation incident upon (i.e. falling onto) a surface at right angles to the solar beam is 7% greater than at aphelion (maximum distance). At its mean distance from the sun, the amount of solar radiation incident at right angles on the outskirts of the atmosphere is 1.396 kWm^{-2}. This is termed the *solar constant.*

3. The altitude of the sun. This depends on the latitude, season and time of day. Generally as the altitude increases the amount of solar radiation incident upon unit area of surface increases.

4. The number of hours of daylight.

5. The transparency of the atmosphere. This together with the above factors dictates the amount of solar radiation incident upon unit area of the earth's surface in a given period. This amount is termed *insolation.*

Assuming the atmosphere has its normal composition and a four eighths cloud cover, the effect of atmospheric transparency is illustrated in Fig. 3.3 and summarized in points A to E below the figure.

The solar radiation incident upon the surface may be described as being either diffuse or direct, indicating whether its passage through the atmosphere has, or has not, been affected by the constituents present. Of the 53 incident units, 6 (F) are reflected by the surface and 47 (G) are absorbed (Fig. 3.3).

The total amount of solar radiation reflected to space (D, E and F) is termed the *planetary albedo,* being the ratio of the solar radiation reflected to that incident on the outer limits of the atmosphere. In this case the planetary albedo is 36%. Its individual components can vary in value. For example if no cloud is present, the 24 units (D) will be incident on the surface, whereas with an extensive cover more than 24 units will be reflected. The albedo value will be influenced by the nature of the surface (forest 5–10%; sand 20–30%; fresh snow 80–90%; sea 7–9%). The angle of incidence of the solar beam upon the surface will also affect the albedo value, which will be of the order of 100% immediately after sunrise and before sunset, and at a minimum when the sun is at the meridian.

In conclusion it should be noted that only 14% (C) of the total amount of incoming solar radiation is directly absorbed by the troposphere. In terms of a gain of energy per unit volume of air the amount is insignificant. In contrast the surface of the earth absorbs a large amount of solar radiation (G).

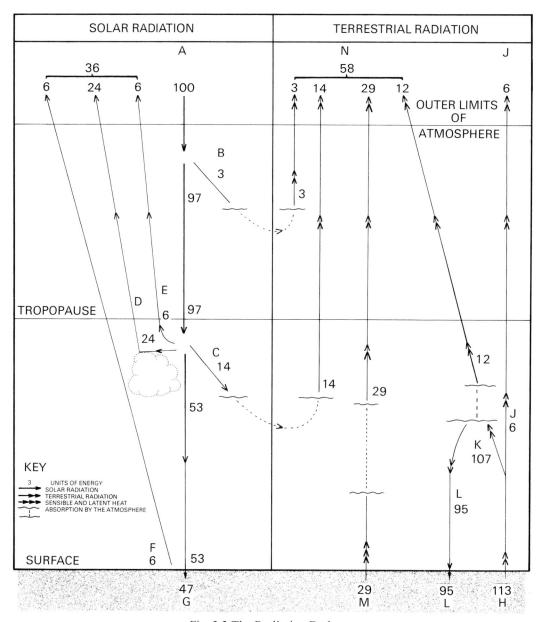

Fig. 3.3 The Radiation Budget.

		Units
A.	The amount of solar radiation incident on the outer limits of the atmosphere.	+100
B.	Absorption in the stratosphere, mainly by ozone, of ultraviolet radiation.	−3
C.	Absorption in the troposphere by gases, water vapour and dust particles.	−14
D.	Reflection to space by clouds, there being no change in the wavelength of the radiation.	−24
E.	Scattering of radiation to space.	−6

(Note: Scattering depends mainly on the ratio of the size of the scattering particle to the wavelength of the radiation incident upon it. In the atmosphere the wavelength usually scattered is that of blue light, hence the blue colouring of the sky.)

Amount of solar radiation incident upon the surface of the earth   53

Terrestrial Radiation and the Atmosphere

While the atmosphere is almost transparent to solar radiation, it is almost opaque to terrestrial radiation absorbing that emitted by the surface and the atmosphere itself. The constituents which play a significant part in the process of absorption are water vapour, carbon dioxide, ozone and cloud, (Fig. 3.3). Assuming the surface has a mean temperature of 288 K (15°C):

		Units
H.	Radiation emitted by the surface.	+113
J.	The part of H which passes through the atmospheric window to space.	−6
K.	The part of H absorbed by the atmosphere.	−107
L.	Radiation emitted by the atmosphere which is incident on the surface and absorbed.	95

Thus the overall energy loss (H−L) from the surface in the form of longwave radiation is 18 units.

On comparing this loss of energy with the gain of 47 units (G), the surface appears to have gained 29 units. However, this energy is transferred to the atmosphere in the form of sensible and latent heat as a result of conduction and evaporation respectively (M). The 58 units of energy which the atmosphere gains through conduction, evaporation and absorption of solar and terrestrial radiation, are eventually radiated to space (N).

The amount of cloud cover and water vapour present will cause variations in energy loss. If the water vapour content is large, or if there is extensive cloud cover at low levels, then a greater part of the longwave radiation emitted by the surface is absorbed by the atmosphere.

In conclusion the atmosphere effectively absorbs terrestrial radiation, but is itself radiating energy continuously. As a large part of the radiant energy emitted by the surface is returned to it, the daily change of air temperature must be attributed to other forms of energy transfer, which are considered below.

ENERGY TRANSFER

The transfer of energy between the surface of the earth and the atmosphere can be attributed to the processes of conduction, convection and evaporation.

Conduction

Conduction is the transfer of energy between two masses which are in contact with each other, from the mass with the higher temperature to that with the lower temperature. The rate at which energy is transferred depends upon the difference of temperature, which is normally expressed in terms of a temperature gradient, and the thermal conductivity of the mass(es). As air is a relatively poor conductor of energy, the process is insignificant within the atmosphere itself. However, where the air is directly in contact with the surface, conduction is important and results in the transfer of energy between atmosphere and surface, depending upon the direction of the temperature gradient.

Convection

Convection is the transfer of energy as a result of movement of parts of a fluid. *Free convection* develops when part of the atmosphere in contact with the surface of the earth is heated through conduction and becomes less dense than the surrounding air. The buoyant mass ascends and eventually redistributes its energy through mixing with air at greater heights. Convection occurs to varying degrees over land during the day, and whenever air passes over a relatively warmer land or sea surface. The terms *convection cells, convection currents* and *convection bubbles* are frequently used to describe the ascending air. The term *thermal* may be used as an alternative in certain contexts.

Forced convection, which is commonly referred to as *turbulence,* develops when air is passing across an uneven surface. Both the degree of turbulence and the height to which it extends increases with increasing wind speed and surface roughness. Vertical transfer of energy occurs within the turbulent layer through eddies, or small pockets of air, which are moving in a random manner. Both free and forced convection may operate simultaneously, while forced convection may be present without the former. Furthermore, forced convection may assist in the exchange of energy between atmosphere and surface in either direction.

Evaporation

Evaporation is the change of a substance from its liquid to vapour state. It commonly occurs at the surface of the earth where liquid water changes to water vapour. The latter entering the atmosphere is a significant source of energy which is released when it condenses. The factors influencing evaporation and the form of energy are discussed in Chapter 4.

DIURNAL VARIATION AND RANGE

During an average 24 hour period the temperatures of the atmosphere and the surface show a systematic change which is termed the *diurnal variation of temperature* (Fig. 3.4). For both land and sea, minimum and maximum surface temperatures are generally attained at sunrise and midday respectively, and the minimum and maximum air temperatures at 1 hour after sunrise and 1500 LMT. In the winter the maximum air temperature occurs at 1400 LMT. The *diurnal range of temperature* (A−B) of both land and sea surfaces is greater than that of the air (C−D) above each respective surface. However, the diurnal range of the surface temperature of the land is greater than that of the sea. For some land surfaces the range may be tens of degrees celsius, whereas for deep sea areas it is less than 1°C. The diurnal range of air temperature above each surface also shows this contrast.

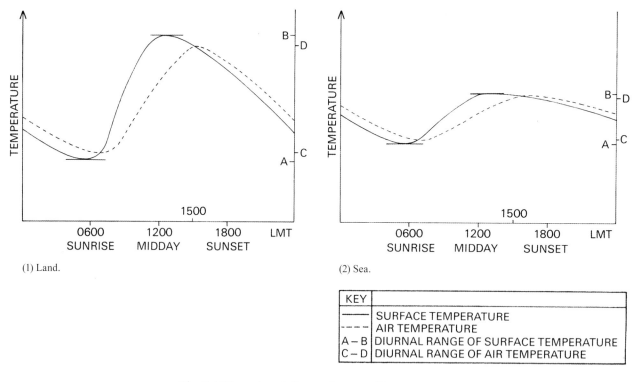

Fig. 3.4 Diurnal variation and range of temperatures.

The diurnal cycle of temperature of a surface and the air above may be analysed in terms of the gains and losses of energy experienced by each medium as follows:

Sunrise to midday.

When the surface temperature increases, the receipt of solar radiation progressively increases to a maximum value at midday. Simultaneously the emission of longwave radiation from the surface increases. Although the surface is absorbing longwave radiation from the atmosphere, the net longwave radiation represents a loss of energy by the surface. However, the solar radiation absorbed during this period more than compensates for this loss.

Midday to sunset.

When the surface temperature decreases, the solar radiation received and the net longwave radiation loss both decrease progressively.

Sunset to sunrise.

The surface temperature continues to decrease but less rapidly. The net longwave radiation loss from the surface continues but in decreasing amounts.

However, the changes in air temperature during the daily cycle (Fig. 3.4) principally depend upon the processes of conduction, convection and turbulence, and thus the influence of the underlying surface.

From one hour after sunrise to 1400/1500 LMT.

The air gains energy through conduction when in direct contact with the surface. Convection and turbulence then ensure the transfer of energy to greater heights.

From 1400/1500 LMT to sunrise.

The temperature gradient between the air and the surface immediately below is such that conduction, aided by turbulence, results in a loss of energy by the air. Its temperature decreases rapidly at first, then more slowly, reflecting the decreasing rate of change of surface temperature during the same period.

Between sunrise and one hour later.

Although the surface experiences a small gain in energy during this period, the air above continues to lose energy, achieving a minimum temperature 1 hour after sunrise.

The significant difference between the values of the diurnal range of land and sea surface temperatures may be attributed to a number of factors:

1. *Specific heat capacity.* The energy required to raise the unit mass of a substance through 1 K is termed *specific heat capacity.* For pure water it is 4.18 J g^{-1} K^{-1}, whereas for any soil mass it is substantially smaller, the absolute value being dependent upon the soil type and its liquid water content.

2. *Transparency to solar radiation.* The depth to which solar radiation penetrates a water mass depends on the amount of solid material contained within it. In pure water the shorter wavelengths within the solar spectrum may penetrate to 100 m before being absorbed, while the longer wavelengths are absorbed by the upper layers. However, solar radiation will only penetrate the first few millimetres of soil. The depth of penetration depends upon the grain size of the soil, longer wavelengths penetrating further than the shorter ones. Since a given amount of solar radiation will be absorbed by a greater mass of water than of land, the increase in water temperature will be correspondingly less.

3. *Evaporation.* The energy absorbed by a surface in the form of solar radiation may be used in the process of evaporation. For a sea surface the amount of energy involved is large, the remaining energy being available to increase the temperature of the water. In contrast the reverse conditions tend to exist for a land surface.

4. *Turbulence.* The generally turbulent nature of sea water aids the distribution of energy to greater depths, thus contributing to its smaller diurnal range of surface temperature compared with that of the land.

To summarise, the small increase in sea surface temperature is a result of its high specific heat capacity value, relative transparency to solar radiation, and the processes of evaporation and turbulence.

The surface temperature of the land decreases more rapidly than that of the sea (Fig. 3.4), since the overall loss of energy experienced by the land surface is only moderated by the relatively small amount of energy gained through conduction from sub-surface levels and the air above. In contrast, when the surface layer of water cools, it becomes denser and sinks, being replaced by less dense warmer water from below. This process, termed *convective overturning,* will occur, provided that the water is at a temperature greater than that of its maximum density, (pure water 4°C). As a result, the sea surface temperature decreases very slowly, but overall there is a loss of energy from the water mass.

Finally there are a number of other factors which affect the value of the diurnal range of temperatures, which are particularly significant for a land area:

1. In mid and high latitudes, the range is greater in the summer due to the longer hours of daylight.

2. Clear skies throughout a 24 hour period result in a greater range than overcast conditions.

3. Advected air, which is non-systematic, also affects air temperature and the diurnal range (Chapter 8).

ENVIRONMENTAL LAPSE RATE

Within the troposphere the air temperature normally decreases with increase in height, with an environmental lapse rate (E.L.R.) of 6.5°C km^{-1} for the standard atmosphere (Chapter 1). However, the E.L.R. value is variable both in time and space. The air temperature may be constant through a limited depth of the troposphere, which is termed an *isothermal layer.* Alternatively, the air temperature may increase as height increases, and this is termed a *temperature inversion* or, more commonly, an *inversion.* Such inversions are either *ground level* if the trend begins at the surface (Fig. 3.5), or *upper level* if it begins at any height above the surface, (Chapter 8). When the air temperature decreases with increase in height, the E.L.R. is positive, whereas a negative E.L.R. exists when the air temperature increases with increase in height.

To establish the E.L.R., upper air temperatures are recorded at 0000 and 1200 UTC by selected stations distributed worldwide, using radiosondes. A radiosonde is a package of instruments attached to a balloon. During ascent the radiosonde monitors and transmits to the surface data on pressure, temperature and humidity. Wind velocity at a number of levels in the atmosphere is calculated using surface radar or Navaids.

The variable nature of the E.L.R., particularly in the lower part of the troposphere, affects the stability of the atmosphere and thus the weather conditions. In Fig. 3.5(1) ELR$_1$ represents the environmental conditions in the lower troposphere. If the air is heated by the underlying surface, the surface air temperature increases (W → X), but the increase in air temperature at greater heights is less (Y → Z). The result is ELR$_2$ which has a steeper (or larger) lapse rate than ELR$_1$. In Fig. 3.5(2), the initial state is that shown by ELR$_3$. If the air is cooled by the underlying surface, the surface air temperature decreases (M → N), but the decrease in air temperature at greater heights is less (R → S). Thus ELR$_4$ is less steep (or smaller) than ELR$_3$. The relative increase/decrease in air temperature at the surface, compared with that at greater heights, underlines the effectiveness of the energy transfer mechanisms between the surface and the atmosphere noted earlier in the chapter.

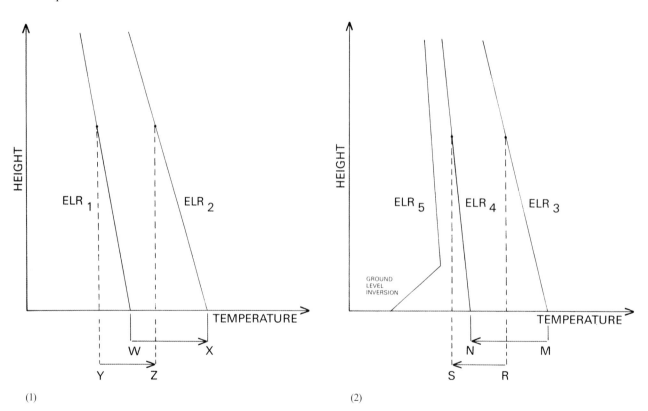

Fig. 3.5 Environmental Lapse Rates.

On occasions, the cooling of air by the underlying surface may result in the development of a ground level inversion ELR$_5$. During the cooling process turbulence plays an important part in aiding the vertical transfer of energy in the atmosphere. However, as the air cools its density increases and turbulence becomes less effective. As a result the mass of air in contact with the surface continues to cool, and its temperature becomes less than that of the air immediately above. The air temperature thus increases with increase in height. Above the inversion the air will experience a relatively small decrease in temperature as a result of the negative value of its net longwave radiation.

In summary, the environmental lapse rate in the lower troposphere increases significantly whenever a mass of air is heated by a land surface during the day, or whenever a mass of air is passing over a relatively warmer land or sea surface. In contrast the environmental lapse rate decreases when air is cooled by the land surface at night, or whenever air is passing over a relatively cooler land or sea surface. Other processes within the troposphere which affect the E.L.R. value are considered in Chapter 8.

CHAPTER 4

WATER IN THE ATMOSPHERE

STATES OF WATER

Water can exist in the atmosphere in the following states:

Vapour – *Water vapour*

Liquid – *Water droplets*

Solid – *Ice crystals*

Of the above, water vapour is invisible, whilst both water droplets and ice crystals are visible. Under certain circumstances it changes from one state to another, and for each process a specific term is used.

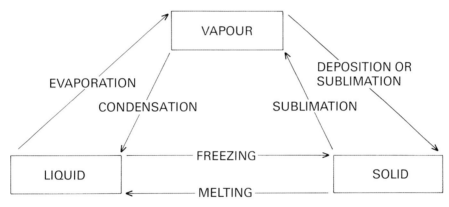

Fig. 4.1 States of water.

The change from liquid to vapour state requires *latent heat of vaporization,* and the change from solid to liquid *latent heat of fusion.* When the changes of state are in the reverse order, latent heat will be released. It should be noted that, when vapour changes directly to the solid state or vice versa, there is no intervening liquid state. *Latent heat of sublimation* is released when vapour changes to solid, and is required when the solid state changes to vapour.

WATER VAPOUR

The quantity of water vapour present in the atmosphere is variable in both time and space. The actual water vapour content of a sample of air may be expressed by a number of terms:

1. *Humidity Mixing Ratio* – the ratio of the mass of water vapour to the mass of dry air (air without water vapour). Units are grams per kilogram (g kg^{-1}).

2. *Absolute Humidity* – the ratio of the mass of water vapour to the volume occupied by the mixture of water vapour and air. This ratio is also known as the *vapour density* or *vapour concentration.* Units are grams per cubic metre (g m^{-3}).

3. *Vapour Pressure* – the pressure exerted by the water vapour in the atmosphere, which forms part of the total atmospheric pressure. Units are hectopascals (hPa).

As its temperature increases, air has the capacity to hold more water vapour. The *saturation curve* (Fig. 4.2) shows the maximum amount which can be present at any given temperature, assuming that the saturated mass of air coexists in equilibrium with a plane liquid water surface. In Fig. 4.2 the vertical axis may be expressed in terms of humidity mixing ratio, absolute humidity or vapour pressure.

Air is termed *saturated* when it contains the maximum amount of water vapour possible at a given temperature. In Fig. 4.2 the mass of air (A) with a temperature of 25°C and a humidity mixing ratio of 21.25 g kg^{-1} is saturated. In this case the humidity mixing ratio is termed the *saturated humidity mixing ratio*. *Unsaturated air,* (B and C) contains less than the maximum amount of water vapour possible at its given temperature. Air is *supersaturated* (D) when it contains more water vapour than is required to saturate it at a given temperature.

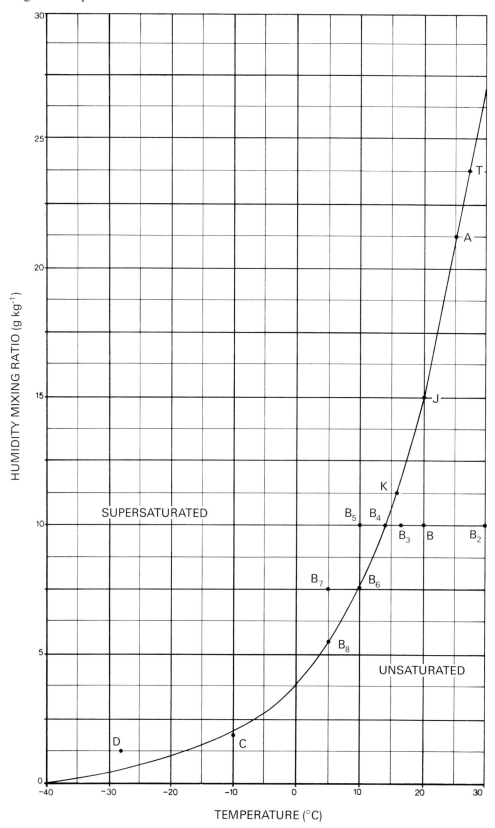

Fig. 4.2 The Saturation Curve.

RELATIVE HUMIDITY

A fourth term most frequently used to describe the water vapour content of the atmosphere is *relative humidity* (RH). This is defined as the ratio of the mass of water vapour present to that which could be present if the air was saturated at the same temperature. The ratio is expressed as a percentage.

The RH value of unsaturated air is always less than 100%. For B at 20°C (Fig. 4.2) the ratio is 10:15 giving an RH value of 66.7% (10/15 × 100), and for C, with the ratio of 1.75:2.00, the RH is 87.5%. It should be noted that although B contains more water vapour compared with C its RH is lower, thus showing that the temperature of the sample, rather than the mass of water vapour, is the controlling factor.

As the temperature of the sample increases or decreases, so the RH value will change. If the water vapour content and pressure are constant and the temperature of B is increased to 30°C (B$_2$), the RH will decrease to 37% (10/27 × 100). If the temperature is decreased to 17°C (B$_3$), the RH will increase to 90.9% (10/11 × 100).

The RH value for saturated air (A) is always 100%, and for supersaturated air (D), which may exist in the atmosphere under certain conditions, it is always greater than 100%.

Changes in relative humidity are not achieved solely by the decrease or increase of air temperature. Saturation can result from an increase in water vapour content (B–J), or by a simultaneous increase of water vapour content and decrease of air temperature (B–K).

DEW-POINT TEMPERATURE

The *dew-point temperature* is the temperature to which a sample of air must be lowered in order to saturate it with respect to a plane liquid water surface, assuming constant pressure and water vapour content. Thus the dew-point temperature of B is 14°C (B$_4$).

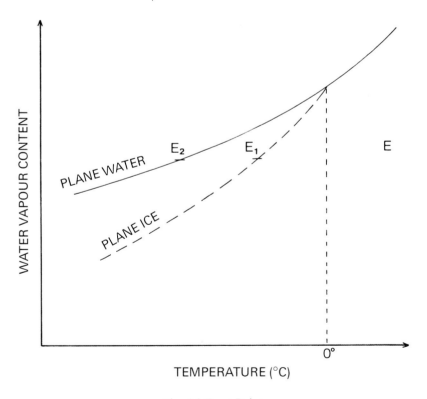

Fig. 4.3 Frost Point.

Below 0°C saturation can be considered with respect to either a plane ice surface or a plane liquid water surface, the latter being supercooled (Fig. 4.3). Assuming constant pressure and water vapour content, the temperature to which a sample of air has to be cooled to become saturated with respect to a plane ice surface will be the *frost point*. In Fig 4.3 for the unsaturated sample of air (E) the value of the frost point (E$_1$) will be greater than that of the dew-point (E$_2$).

CONDENSATION

Condensation of water vapour in the atmosphere is common, and occurs usually as a result of the decrease of air temperature below its dew-point temperature. In Fig. 4.2, B_4 with a dew-point temperature of 14°C is decreased to 10°C (B_5) and becomes supersaturated. This state is rarely maintained in the atmosphere because some water vapour condenses ($B_5 - B_6$), and the air becomes saturated (B_6). Further cooling of the air ($B_6 - B_7$) would result in further condensation ($B_7 - B_8$).

In the atmosphere condensation requires the presence of condensation nuclei, which are minute particles present in varying concentrations. These have originated from the surface of the earth as a result of natural processes (volcanic activity, bursting of air bubbles from the sea), from man's activity, and from chemical processes occurring in the atmosphere itself. Sodium chloride (common salt) and sulphuric acid are examples of condensation nuclei and are hygroscopic (have an affinity with water). The nuclei become part of the water droplets and maintain the existence of the latter in saturated air. Water droplets take varying forms in the atmosphere, depending upon the process which has caused the decrease in the temperature (Chapter 5). In theory it is possible for a mass of air cooled below its dew-point temperature to remain supersaturated (D), provided there are no condensation nuclei present.

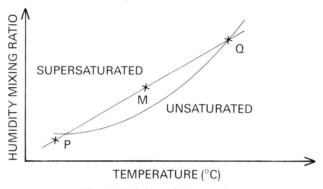

Fig. 4.4 Mixing of air samples.

The mixing of two samples of air having different temperatures and RH values may result in condensation. In Fig. 4.4, the mixing of Q whose RH is 100%, and P at less than 100% will result in M, whose value of water vapour content and temperature will be between those of P and Q. M is supersaturated and condensation occurs provided that nuclei are present.

EVAPORATION

The water vapour content of the atmosphere is derived through the process of evaporation from the surface of the earth, and may also be the result of sublimation where there is ice and snow. Sources of water, from which evaporation occurs, include not only free water surfaces (oceans, lakes and rivers), but also soil and vegetation. Evaporation is expressed as a rate, being the depth of liquid water which changes into water vapour in a given period. Over land it is termed the *evapotranspiration rate* since it includes evaporation from soil and transpiration from vegetation.

The rate of evaporation depends upon a number of factors:

1. Energy. Evaporation requires latent heat of vaporization and the energy for this process is ultimately derived from solar radiation absorbed by the surface.

2. Relative Humidity. In Fig. 4.2, when the air immediately above the surface is saturated (A), the evaporation rate will be zero, since no more water vapour can be absorbed. If the air is unsaturated (B) then evaporation occurs.

3. Wind. When saturated air (A) is replaced through air movement (i.e. wind) by unsaturated air (B), the evaporation rate increases. Evaporation can continue if the air at the surface once saturated, is replaced by unsaturated air from a higher level due to turbulence.

4. Water. All other conditions may be favourable for evaporation, but lack of water, whether free surface or from vegetation, will prevent the optimum rate being achieved. Such conditions occur in deserts and areas with dry seasons.

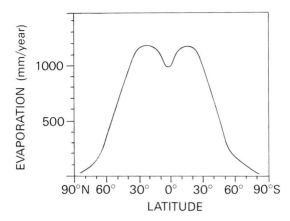

Fig. 4.5 Average annual evaporation distribution for ocean areas.

The annual distribution of evaporation for ocean areas is illustrated in Fig. 4.5. The most favourable conditions for evaporation in terms of water, energy availability, and atmospheric conditions, exist in latitudes 15–30°, and the least favourable conditions over polar regions. The result of evaporation is the transfer of energy from the surface of the earth to the atmosphere.

DIURNAL VARIATION OF RELATIVE HUMIDITY

Relative humidity at any observing station generally varies throughout the day, with a minimum value in mid-afternoon and a maximum in the period before dawn (Fig. 4.6). However, the diurnal range depends upon the nature of the surface, the season, and the temperature and water vapour content of the air. The diurnal range is generally greater over land than sea.

If the water vapour content of the air remains constant, the RH value decreases during the day as air temperature increases (Fig. 4.2 (B–B$_2$)). The RH will therefore be at a minimum when air temperature is at a maximum (Chapter 3). Thereafter the RH increases as air temperature decreases, and the air may reach its dew-point temperature. If the RH reaches 100%, it will remain at this value until the air temperature begins to increase 1 hour after sunrise.

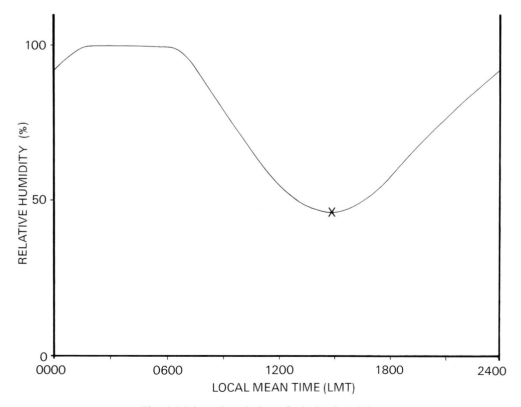

Fig. 4.6 Diurnal variation of relative humidity.

As the air temperature increases during the day, the air is capable of holding an increasing amount of water vapour, and evaporation occurs if conditions are favourable. However, the increase in air temperature has a greater influence on the RH value than the increase in water vapour content. There is therefore an overall decrease in RH value during the period. If, during the subsequent cooling period, the air temperature decreases below its dew-point temperature, the RH remains at 100% and condensation occurs (Fig. 4.2 ($B_5-B_6-B_7-B_8$)). This explanation has considered only the immediate surface and the air mass above it. However, it should be noted that the movement of air from other areas will also introduce irregularities into the diurnal variation of relative humidity (Chapter 7 and 8).

HYGROMETERS

Hygrometers have been developed to assist in determining the water vapour content of air. The most commonly used hygrometer is *Mason's Hygrometer,* a dry and a *wet-bulb thermometer* which are housed in the Stevenson or Marine Screen (Chapter 3). A second form is the *whirling psychrometer,* or hygrometer, in which the dry and wet-bulb thermometers are fixed in a framework which can be rotated by hand (Fig. 4.7). The wet-bulb thermometer has its bulb encased in muslin. This is kept damp with distilled water drawn by a wick from a reservoir. Where a platinum resistance thermometer is used this is also encased in muslin.

The working principle of the wet-bulb thermometer is that, when ventilated with unsaturated air, water evaporates from the damp muslin. Due to the latent heat of vaporization required for the change of state, the wet-bulb thermometer will record a lower temperature compared to the dry-bulb thermometer. The difference between the readings on the dry- and wet-bulb thermometers is called the *depression of the wet bulb,* and the lower the RH value the greater will be the depression. If the air has an RH of 100% there will be no depression.

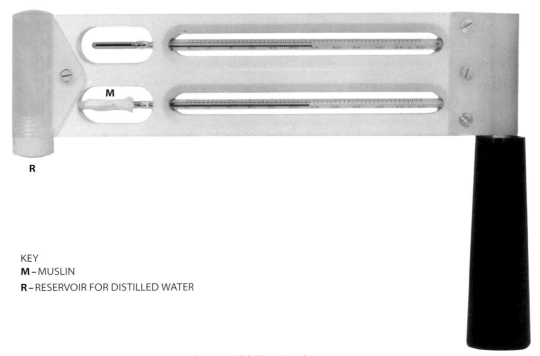

KEY
M – MUSLIN
R – RESERVOIR FOR DISTILLED WATER

Fig. 4.7 Whirling Psychrometer.

The rate of evaporation of water from the wet-bulb thermometer also depends on the rate of air flow, being 2–4 knots for the Mason's Hygrometer, and at least 7 knots for the whirling psychrometer. Each instrument therefore requires a separate set of hygrometric tables. These tables give values for dew-point temperature, relative humidity, and vapour pressure for each reading of the dry-bulb temperature and depression of the wet-bulb.

It should be noted that, if the wet-bulb reading is below 0°C, it is assumed that the muslin is coated with ice. If supercooled water (water in liquid form at a temperature below 0°C) exists on the bulb, it must be converted to a coating of ice, and the instrument ventilated before a reading is taken.

CHAPTER 5

CLOUDS

INTRODUCTION

Clouds are collections of water droplets or ice crystals, or combination of these two states of water, suspended in the atmosphere. A knowledge of the many types of clouds and their occurrence provide a valuable source of information to the seafarer in forecasting the weather.

CLOUD TYPES

The shapes of cloud within the troposphere may be *stratiform* (flattened or layered), *cumuliform* (heaped), *cirriform* (hair or thread-like), or a combination of these. There are ten basic genera or characteristic forms, and further subdivision into species and varieties can be made. The internationally agreed classification of the ten genera is related to the height of the cloud base above the surface (Table 5.1). Table 5.2 on page 32 sets out the salient features of each genera.

Cloud base	Genera	Abbreviation	Height of base in kilometres		
			Tropics	Mid lats.	High lats.
HIGH	Cirrus	Ci	>6	>5	>3
	Cirrostratus	Cs			
	Cirrocumulus	Cc			
MEDIUM	Altostratus	As	2 − 7.5	2 − 7	2 − 4
	Altocumulus	Ac			
LOW	Stratus	St	<2	<2	<2
	Stratocumulus	Sc			
	Nimbostratus	Ns			
	Cumulus	Cu			
	Cumulonimbus	Cb			

Table 5.1 Cloud genera.

Alto identifies the medium level clouds, and *nimbus* implies rain, but other forms of precipitation are possible.

High level cirriform clouds are a collection of ice crystals. For medium and low level clouds the composition will be dependent on the temperature distribution through the vertical extent of the cloud.

ADIABATIC LAPSE RATE

Cloud formation is mainly the result of air ascending and cooling adiabatically. When a parcel of air ascends, the pressure exerted on it by the surrounding atmosphere decreases, so allowing the parcel to expand. In order to do so it requires energy which is derived from the parcel itself, and its temperature therefore decreases. Since air is a poor conductor, it is assumed that no energy is exchanged between the air parcel and the surrounding atmosphere. This process, in which no heat enters or leaves the system, is termed adiabatic from the Greek word meaning "impassable". When an air parcel descends, the reverse process occurs and its temperature increases. The rate at which the temperature of the parcel changes with height is termed the *Adiabatic Lapse Rate*. For a *dry air parcel*, in which the air is unsaturated, the rate is 9.8°C km^{-1} (usually rounded up to 10.0°C km^{-1}). This is the *Dry Adiabatic Lapse Rate* (D.A.L.R.), which is applicable whether the air parcel is ascending or descending.

An ascending *saturated air parcel* will cool at the *Saturated Adiabatic Lapse Rate* (S.A.L.R.), the value of which is less than the D.A.L.R.. During its ascent the volume of the air parcel increases, and its temperature decreases, as for an unsaturated air parcel. As a result some water vapour condenses, releasing latent heat of vaporization. This energy causes an increase in the temperature of the air parcel, which partly compensates for the energy used during the process of expansion. Thus, during its ascent the temperature of the parcel will decrease, but the amount will depend on the quantity of water vapour condensing (Fig. 5.1). In the example, A_1 at 20°C cools to 15°C (A_2) and 4.5 g of water vapour condense. B_1 at 0°C cools to -5°C (B_2) and 1.25 g of water vapour condense. Thus the gain in energy as a result of condensation is greater in the case of A than B, and the overall rate of decrease in the temperature of A will be less than that of B. Thus the S.A.L.R. is variable and its value may be between 3°C km^{-1} and 9°C km^{-1}.

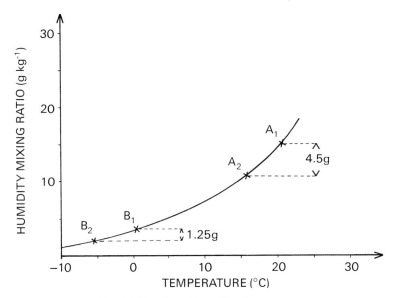

Fig. 5.1 Condensation of water vapour.

ATMOSPHERIC STABILITY

An assessment of the stability of the atmosphere based on a knowledge of the Environmental, Dry and Saturated Adiabatic Lapse Rates is of considerable value in forecasting cloud development and related weather conditions.

The atmosphere is *absolutely unstable* when a saturated or unsaturated air parcel, ascending and cooling adiabatically, has a tendency to continue its displacement. In Fig. 5.2(1), an air parcel at level AA has a temperature greater than that of the surrounding atmosphere. It is therefore less dense and, being buoyant, continues to ascend. For a descending parcel, warming adiabatically, its temperature is less than that of the surrounding atmosphere at any given level, and the parcel continues to descend. In an absolutely unstable atmosphere, the E.L.R. is greater than D.A.L.R., which is greater than S.A.L.R. (E.L.R.>D.A.L.R.>S.A.L.R.).

The atmosphere is *absolutely stable* when any saturated or unsaturated air parcel, ascending and cooling adiabatically, has a tendency to return to its original level. In Fig. 5.2(2), the temperature of an air parcel at level BB is less than that of the surrounding atmosphere. The parcel is therefore denser and tends to return to its original level. A descending air parcel, warming adiabatically, tends to return to its original level, since it is warmer than the surrounding atmosphere. In an absolutely stable atmosphere, the D.A.L.R. is greater than S.A.L.R., which is greater than E.L.R. (D.A.L.R.>S.A.L.R.>E.L.R.).

The atmosphere is *conditionally unstable* when an unsaturated air parcel, ascending and cooling adiabatically, is at a lower temperature than the surrounding atmosphere at any level. In Fig. 5.2(3), this condition exists at level CC, and the air parcel tends to return to its original level, since it is denser than the surrounding atmosphere. However, at this level a saturated air parcel, which has ascended and cooled adiabatically, has a temperature greater than that of the surrounding atmosphere and continues to ascend. In a conditionally unstable atmosphere, the D.A.L.R. is greater than E.L.R., which is greater than S.A.L.R. (D.A.L.R.>E.L.R.>S.A.L.R.).

The atmosphere is in a state of *neutral equilibrium* when the E.L.R. equals D.A.L.R., or the E.L.R. equals S.A.L.R.. In each case the air parcel, ascending (descending) and cooling (warming) adiabatically, remains at its new level.

The D.A.L.R. is always constant and, if the S.A.L.R. is assumed to have a fixed value, the E.L.R. will be critical in determining the stability of the atmosphere at a given time. By recording air temperature at increasing heights during a radiosonde ascent, it is possible to establish the E.L.R. for different layers in the troposphere, and thus assess the stability in each of these layers.

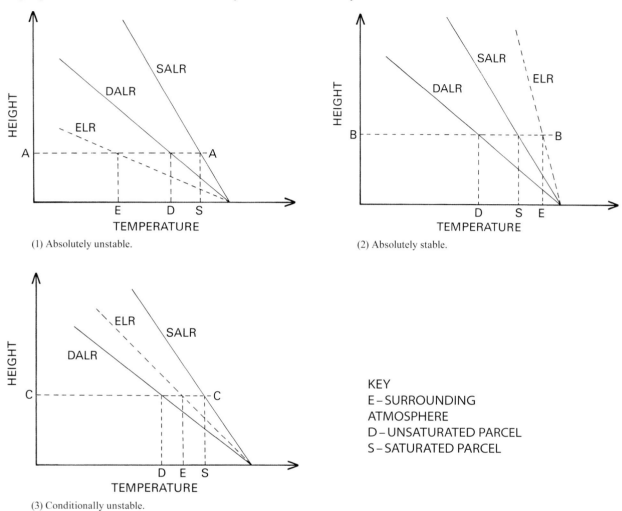

(1) Absolutely unstable.

(2) Absolutely stable.

KEY
E – SURROUNDING
ATMOSPHERE
D – UNSATURATED PARCEL
S – SATURATED PARCEL

(3) Conditionally unstable.

Fig. 5.2 Atmospheric stability.

FORMATION OF CLOUDS

The stability of the atmosphere plays an important part in the formation and development of clouds and their characteristics, since it controls the method of ascent of an air parcel.

Convection

A convection current, developing at the surface, will ascend cooling adiabatically, and continue to do so as long as it is warmer than the environment (Fig. 5.3(1)). If the air parcel becomes saturated during its ascent it cools at the S.A.L.R.. Water vapour condensing forms water droplets, which are visible in the shape of cumuliform cloud. The base of the cloud is at the condensation level, being the height at which the convection current begins to cool at the S.A.L.R.. The dimensions of the cloud base partly reflect the dimensions of the convection current, whilst the vertical extent of the cloud depends upon the height to which the current ascends. In these circumstances the atmosphere immediately above the surface is absolutely unstable (E.L.R.>D.A.L.R.>S.A.L.R.), and this condition is termed *superadiabatic,* and results from the heating of the atmosphere by the surface (Chapter 3).

Cumuliform cloud may be fair-weather cumulus (Plate 9) with limited vertical extent (Fig. 5.3(1)), towering cumulus (Plate 10: Fig. 5.3(2)), or cumulonimbus (Plate 11: Fig. 5.3(3)). A distinctive contrast exists between the first two and cumulonimbus, as the former have sharp outlines, while the latter has a smooth fibrous or striated upper portion, described as diffuse. A sharp outline denotes a cloud composed of water droplets or supercooled water droplets, which on the periphery of the cloud evaporate into the surrounding unsaturated air. The energy required for evaporation results in a decrease in temperature of the air, which becomes denser and sinks to produce the well-defined edge to the cloud.

Cumulonimbus normally develops from towering cumulus, and the change to a diffuse outline is observed when the water droplets on the upper edge of the cloud change into ice crystals (the cloud has become glaciated). An anvil is formed as the crystals tend to drift into the surrounding atmosphere where they sublimate slowly (Plates 12, 13 and 14).

If there is an upper level inversion (Fig. 5.3(4)), cumulus cloud ceases development vertically at this level, but may spread horizontally, forming stratocumulus or altocumulus depending on the height at which the inversion exists (Plate 15).

The development of a cumuliform cloud depends upon a stream of convection currents from the surface. The air parcel concept assumes that the currents retain their identity during ascent, but this is not the case since environmental air will mix with the convection currents, a process called *entrainment*. This process tends to prevent the initial currents reaching condensation level, but modifies the environmental air through which successive currents ascend, thus allowing the latter to reach greater heights.

The variation of wind direction and speed with height, termed *vertical wind shear,* may also affect the shape of cumuliform clouds, which may appear to be either more advanced or retarded with increasing height.

On occasions convection currents may generate a funnel shaped cloud, termed a *waterspout,* which may exist for up to half an hour (Plate 16). A waterspout develops at the base of a cumulonimbus cloud, from which it descends to the surface, and may be appreciably bent by vertical wind shear. Its diameter may vary from a few metres to a few hundred metres and within this area it generates confused seas.

Orographic

Air moving across the surface of the earth will be forced to ascend over a hill or mountain lying in its path, and thus cools adiabatically. The clouds which may form are called *orographic* and are associated with a stable atmosphere, where air has to be forced to ascend (Fig. 5.3(5)). Cloud may be stratus or nimbostratus, with a base at the condensation level and an upper limit depending upon the height of the barrier (Plate 17). The Tablecloth, a well known example, is a stratus cloud which forms on the windward side of Table Mountain in South Africa.

Dependent upon the distribution of temperature, wind direction and speed with height, a wave motion may be generated downwind of a physical barrier. This condition can be observed by the presence of lens-shaped or lenticular clouds, which indicate where the air is rising on each successive wave crest (Plate 18). These clouds are often visible at substantial distances downwind of the barrier.

In a conditionally unstable atmosphere orographic conditions play an important part in forcing an unsaturated air parcel to ascend (Fig. 5.3(6)). The parcel will cool adiabatically and become saturated, and further forced ascent and cooling at the S.A.L.R. may eventually result in the air parcel (Y) having a temperature greater than that of the surrounding air. The parcel will then rise spontaneously until its temperature is the same as that of the surrounding atmosphere. Cumulus or cumulonimbus cloud develops.

Turbulence

Turbulence, by redistributing the water vapour uniformly and affecting the E.L.R. within the turbulent layer, can cause cloud formation (Chapter 3). Above a certain height within the layer, condensation occurs and stratus or stratocumulus cloud develops (Fig. 5.3(7)). Turbulence is therefore another process which can force air to ascend in a stable atmosphere. Although it is only effective to a limited height, the horizontal area covered by the cloud is usually extensive over land and sea.

Frontal

Cloud development occurs within frontal depressions, where adiabatic cooling of the large scale vertical motion of warm air in the troposphere results in the condensation of water vapour. Details of cloud types associated with various fronts may be found in Chapter 8.

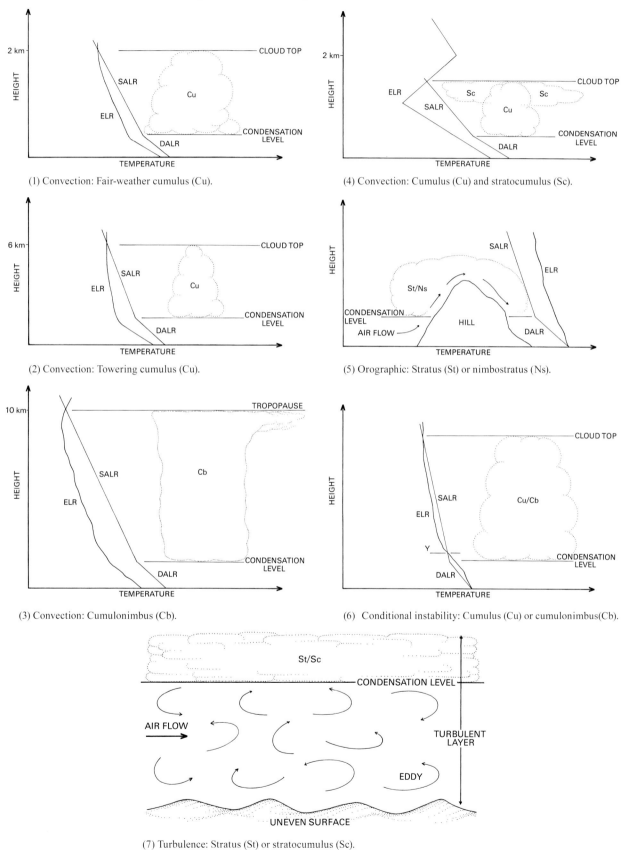

Fig. 5.3 Formation of clouds.

		Plate
	High Clouds	
Cirrus (Ci)	Detached clouds in the form of white delicate filaments or white or mostly white patches or narrow bands. These clouds have a fibrous appearance or a silky sheen or both.	1
Cirrostratus (Cs)	A transparent whitish cloud veil of fibrous appearance or smooth appearance totally or partly covering the sky, and generally producing halo phenomena.	2
Cirrocumulus (Cc)	A thin, white patch, sheet or layer of cloud without shading, composed of very small elements in forms of grains or ripples merged or separate and more or less regularly arranged.	3
	Medium Clouds	
Altostratus (As)	A greyish or bluish cloud or layer of striated, fibrous or uniform appearance, totally or partly covering the sky, and having parts thin enough to reveal the sun at least vaguely.	4
Altocumulus (Ac)	A white or grey, or both white and grey, patch, sheet or layer of cloud, generally composed of rounded masses or rolls, which are sometimes partly fibrous or diffuse and which may not be merged.	5
	Low Clouds	
Stratus (St)	A generally grey cloud layer with a fairly uniform base. When the sun is visible through the cloud its outline is clearly discernible.	6
Stratocumulus (Sc)	A grey or whitish or both grey and whitish patch, sheet or layer of cloud which almost always has a dark part, composed of rounded masses or rolls, which are non-fibrous, and which may or may not be merged.	7
Nimbostratus (Ns)	A grey cloud layer, often dark, whose appearance is rendered diffuse by more or less continuously falling rain or snow, which in most cases reaches the ground. It is thick enough throughout to blot out the sun.	8
Cumulus (Cu)	Detached clouds, generally dense and with sharp outlines, developing vertically in the forms of rising mounds, domes or towers, of which the bulging upper part often resembles a cauliflower. The sunlit parts of these clouds are mostly brilliant white and their bases relatively dark and nearly horizontal.	9 10
Cumulonimbus (Cb)	A heavy dense cloud, with a considerable vertical extent, in the form of a mountain or huge towers. At least part of its upper portion is usually smooth, fibrous or striated, and nearly flattened; this often spreads out in the form of an anvil or vast plume.	11

Table 5.2 Description of cloud genera.

CHAPTER 6

PRECIPITATION AND FOG

FORMS OF PRECIPITATION

Precipitation is the deposit on the earth's surface of water in liquid or solid state or a combination of both. The principal forms are:

Drizzle – Water droplets with diameters between 200 μm and 500 μm.

Rain – Water droplets with diameters exceeding 500 μm.

Snow or Snowflakes – Small ice crystals or aggregates of ice crystals.

Hail – Balls of ice of varying size.

Sleet – Mixture of rain and snow.

Ice pellets, prisms or granular snow also occur. Usually, but not always, precipitation is associated with a cloud. On occasions it can be seen leaving the base of a cloud in vertical or inclined trails which do not reach the surface, which are termed *fallstreaks* or *virga*.

DEVELOPMENT

Cloud droplets have diameters of the order of 20 μm compared with drizzle droplets of 200 μm or more. Investigations have established that the increase in size of droplets is not caused by condensation within the cloud. In a cloud composed entirely of droplets whose temperatures are greater than 0°C the theory of

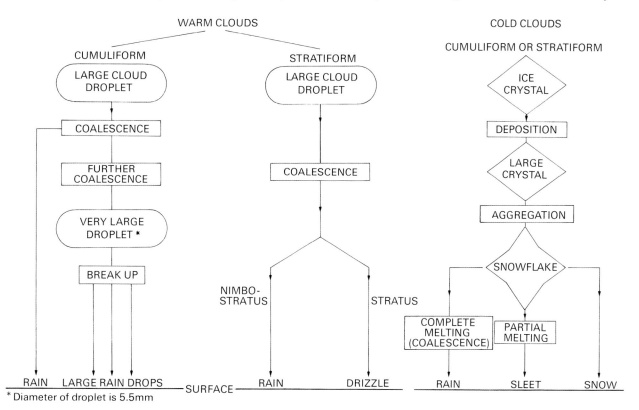

Fig. 6.1 Development of precipitation.

coalescence applies. The size of a cloud droplet is directly related to the size of the condensation nucleus on which it forms. A large droplet has a greater speed of descent compared with a smaller one, and it is therefore possible for the former to collide with the latter lying in its path and they coalesce (join together).

As the larger droplet gathers momentum further collision and coalescence occurs. Eventually the droplet leaves the cloud when its size is such that its speed of descent exceeds the upward movement of air, or *updraught,* within the cloud. The droplet may decrease in size through evaporation before it reaches the surface, depending upon the relative humidity of the air below the cloud base.

The clouds in which this process occurs are termed "warm" clouds, and the type of cloud determines whether drizzle or rain is experienced. For example, stratus produces drizzle, whereas nimbostratus and cumuliform clouds produce rain (Fig. 6.1). Rain drops from cumulus and cumulonimbus clouds can be very large as a result of the concentration of water droplets within the cloud, and the increased frequency of collision due to smaller droplets being lifted by the updraught.

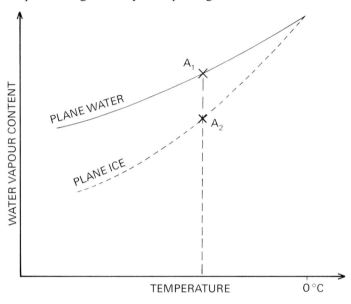

Fig. 6.2 Ice crystal development.

In a "cold" cloud, in which the air temperature is less than 0°C in part if not throughout the cloud, the Bergeron–Findeisen theory applies. In these clouds the water droplets are supercooled. The development of ice crystals from these droplets depends upon the presence of freezing nuclei, which have a crystalline structure similar to that of the hexagonal form of ice. Many of these nuclei are introduced into the atmosphere from the surface, and each type of nucleus has a threshold temperature below which a supercooled water droplet will freeze on coming into contact with it. An important nucleus is the clay mineral kaolinite whose threshold temperature is −9°C. With decreasing air temperature different types of freezing nuclei become active, and at −22°C and below a cloud is composed entirely of ice crystals.

A newly formed ice crystal is surrounded by air (Fig. 6.2 (A_1)) which is saturated with respect to a plane water surface, and supersaturated with respect to a plane ice surface. Water vapour will therefore sublimate directly onto the ice crystal which increases in size. The surrounding air then becomes unsaturated with respect to a plane water surface (A_2), thus the supercooled water droplets present decrease in size through evaporation. As this process is repeated the crystal grows. Air currents within the cloud may cause the crystal to fragment, and each fragment then forms a further nucleus, thus increasing the number of crystals within the cloud. Collision and aggregation (joining together) of ice crystals may occur, particularly between 0°C and −5°C, resulting in large snowflakes.

When the snowflake is of such a size that its speed of descent overcomes the updraught it will leave the base of the cloud. Snow occurs when the surface air temperature is below 0°C, but between 0°C and 3°C some of the snowflakes melt, resulting in sleet (Fig. 6.1). When an ice crystal descends into the lower part of the cloud where the temperature is greater than 0°C, it melts forming a water droplet which then grows through collision and coalescence, and reaches the surface in the form of rain. These processes can occur in cumulus, cumulonimbus and nimbostratus clouds (Fig. 6.1).

Fig. 6.3 Hailstone. Cross section showing alternating clear and opaque ice.

Hailstones

Hailstones are associated with cumulonimbus clouds, and are usually spherical, with diameters ranging from 5 mm to 50 mm or more. The structure of each stone is a series of concentric shells of alternating clear and opaque ice (Fig. 6.3), which is the result of low and high air concentrations respectively. In a cumulonimbus cloud the strong updraught lifts water droplets into its upper part where they freeze, and become the nuclei of hailstones. Each hailstone may then increase in size through collision and accretion (joining together) with water droplets. The developing hailstone then descends and, if it remains within the cloud, rejoins the updraught. This process may be repeated several times until its speed of descent overcomes the updraught and it leaves the cloud (Fig. 6.4).

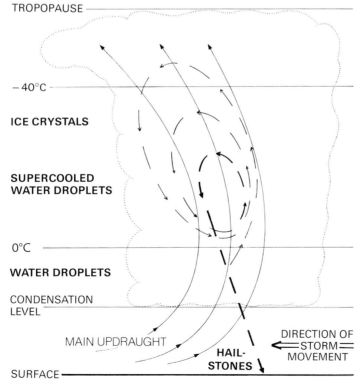

Fig. 6.4 Development of hailstones.

OBSERVATION

A number of terms are used for observing purposes to describe the precipitation reaching the surface:

Showers – From convective clouds of short duration in the form of rain, snow, hail or sleet with rapid fluctuation of intensity (Plate 19).

Intermittent precipitation – From stratiform clouds, when there are breaks in the precipitation within the past hour.

Continuous precipitation – From stratiform clouds, when it has lasted for at least an hour without a break.

Precipitation whether from cumuliform or stratiform clouds may be described as *slight, moderate* or *heavy.* Each term indicates the amount, as a depth in millimetres, reaching the surface during one hour. Particular values are noted in observers' handbooks.

Thunderstorms, which are flashes of lightning from electrical discharges and thunder, may be associated with very heavy, or violent, showers of hail or rain from cumulonimbus cloud (Plate 20). Simultaneously there is an increase in surface wind speed and pressure, and a decrease in air temperature, which together reflect the *downdraught* developed in the cloud as a result of the precipitation.

Drizzle or rain may freeze on coming into contact with the surface of the earth or any object standing above it, when it is termed *freezing drizzle or freezing rain.* The ice formed is termed *glazed frost,* or *black ice* when encountered on roads, and at sea, *icing* or *ice accretion.*

VISIBILITY

For meteorological purposes horizontal visibility is defined as the greatest distance at which an object with specified characteristics can be seen and identified by the unaided eye in daylight. At night it is assumed that the illumination of the object is raised to normal daylight level.

Visibility is assessed by viewing the horizon through 360° and recording the shortest distance. Land observing stations use objects at known distances in daytime and a visibility meter at night, thus making it possible to provide accurate visibility ranges:

Visibility	*Scale*
If less than 5 km	0.1 km steps
If between 5 & 30 km	1 km steps
If over 30 km	5 km steps

Table 6.1 Visibility scale used on land.

At sea the limited availability of objects often makes the estimation of visibility difficult and a coarser scale is used:

Range Recorded in Steps			
km	*nautical mile*	*km*	*nautical mile*
<0.05	<0.03	2.0 – 4.0	1.1 – 2.2
0.05 – 0.2	0.03 – 0.1	4.0 – 10.0	2.2 – 5.4
0.2 – 0.5	0.1 – 0.3	10.0 – 20.0	5.4 – 11.0
0.5 – 1.0	0.3 – 0.5	20.0 – 50.0	11.0 – 27.0
1.0 – 2.0	0.5 – 1.1	≥50.0	≥27.0

Note 1. ≥ means "greater than or equal to": < means "less than".
Note 2. This table is based on the international scale using km.

Table 6.2 Visibility scale used at sea.

Visibility is reduced by the suspension of liquid or solid particles in the atmosphere. If the visibility is reduced to less than 1 km as a result of water droplets, the condition is termed *fog,* and if 1 km or greater it is termed *mist.*

If visibility is reduced by the presence of solid particles the condition is termed *haze,* for which there is no upper limit to the value of the visibility range. When dust and sand are lifted into the atmosphere resulting in a visibility range of less than 1 km, the condition is termed a *duststorm* or *sandstorm,* and above this range a *dust* or *sand haze*. Other natural sources of reduced visibility are volcanic activity resulting in ash and forest fires. Notable reductions in horizontal visibility can be due to anthropogenic activity, sources being industrial and vehicular pollution and fires associated with land clearance. Although generated on land the effects can extend offshore as noted off the east and south-east coasts of Asia.

FOG

Fog develops for a variety of reasons and a number of types can be identified:

Advection fog

Sea Smoke

Radiation fog

Frontal fog (see Chapter 8)

Advection Fog

Advection fog develops as a result of a mass of warm air, with a high relative humidity value, moving horizontally (hence the term *advection*) over a cooler surface, whose temperature is below the dew-point temperature of the air. When, as a result of conduction aided by turbulence, the air is cooled below its dew-point temperature, water vapour condenses, the water droplets producing the mist/fog condition.

This type of fog forms and persists under a wide range of wind speeds. The degree of turbulence dictates the maximum height to which the air is cooled, the height increasing with increasing wind speed. Thus the temperature gradient between air and surface in conjunction with the degree of turbulence determines the likelihood of advection fog. Low wind speeds provide more favourable conditions but, with a steep temperature gradient, advection fog can develop in gale force winds. However, higher wind speeds associated with small temperature gradients are more likely to produce low level stratus cloud, as the effective cooling of the air by the surface is less and is spread over a greater height. The lower troposphere normally becomes stable since cooling of the air decreases the environmental lapse rate.

At sea *advection fog,* termed *sea fog,* occurs at certain times of the year (Fig. 6.5). In northern latitudes, the Grand Banks of Newfoundland and the North Pacific zones are notorious particularly in July, when warm air from the south-west and south pass over the cold waters of the Labrador, and the Oyo Shio or Aleutian Currents respectively. Sea fog in these areas can persist for extended periods and will only disperse when either the wind speed increases, or its direction changes. Sea fog also occurs in lower latitudes during the summer in the region of the cold California, Canary, Peru and Benguela Currents.

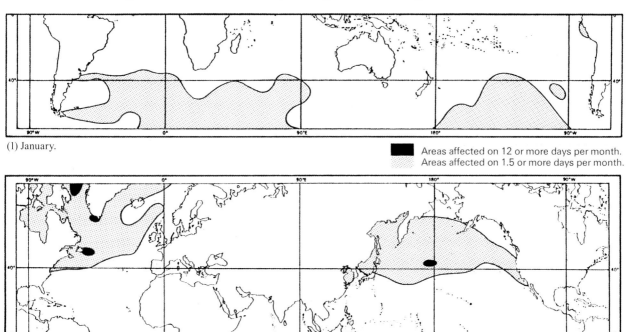

(1) January.

Areas affected on 12 or more days per month.
Areas affected on 1.5 or more days per month.

(2) July.

Fig. 6.5 Distribution of sea fog.

Sea fog not only develops where cold currents exist, but also where there are favourable conditions of wind speed, air and sea surface temperatures. Examples are the spring and early summer fogs of the Western Approaches to the British Isles, where the south-westerly warm air stream from the Azores moves over the sea which, at this time of the year, is at its lowest temperature (Plates 21, 22 and 23). In the North Sea, sea fog develops during the summer when warm north-east, east and sometimes south-easterly winds from Europe pass over the colder sea surface. Along the east coast of the British Isles this sea fog is called *haar* or *sea fret.*

On land, warm air moving over cold surfaces may also produce advection fog. In the British Isles this usually occurs in winter through advection of a warm air stream from the Azores. At this time of year advection fog also develops over the southern and eastern areas of the United States of America, when warm air is advected from the Gulf of Mexico and the Bermuda region.

Sea fog is a frequent threat to the seafarer and its prediction is therefore important. As sea and dew-point temperatures are critical in its formation, their observation at frequent intervals is recommended, and should be recorded in graphical form (Fig. 6.6). By drawing straight lines to establish the trend of each temperature, it is possible to determine the point of intersection and thus the time (F), when fog may be encountered.

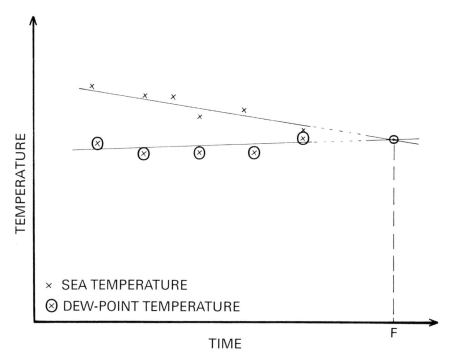

Fig. 6.6 Forecasting sea fog.

Sea Smoke

Sea smoke, arctic sea smoke, frost smoke, or steam fog is present when the surface of the sea has a steaming or smoky appearance. This fog is often patchy and extends to a limited height above the surface, with good visibility at bridge level but poor from the upper deck. The condition is caused by the movement of cold air over a warmer surface, the temperature difference usually being of the order of 10°C, although given favourable wind conditions it may occur with smaller differences. The air immediately above the surface is heated and becomes saturated through evaporation from the surface. It ascends and mixes with colder unsaturated air above (Fig. 4.4). Since the mixture is supersaturated, condensation occurs and the water droplets form sea smoke. As the air is heated by the underlying surface, the environmental lapse rate of the lower troposphere will be that of an unstable atmosphere.

The wind speed associated with the formation of sea smoke may vary from very low to gale force. Higher speeds are more favourable when the temperature difference is small, as they ensure a continuous supply of cold air immediately above the surface.

Off the east coasts of the North American and Asian continents sea smoke occurs during the winter months, when cold air from the continent passes over estuaries, coastal waters, (e.g. St. Lawrence Seaway – Plate 24) and adjacent ocean areas. During winter it occurs in the Baltic Sea which is surrounded by a colder land mass, and in higher latitudes it is associated with cold winds from the Arctic Basin and the ice covered sea areas to the south. In lower latitudes it occurs occasionally in the Gulf of Mexico and off Hong Kong.

As the air temperature is low, the droplets forming sea smoke are often supercooled, and icing occurs when the droplets freeze on contact with those parts of a vessel which are below 0°C.

Radiation Fog

Radiation fog is a land based fog in its development. Clear skies, a high relative humidity, very low wind speeds, and a relatively long period during which the air can cool, are the most suitable conditions for its formation. The clear sky condition allows the maximum loss of long wave radiation from the surface during the night. Surface temperatures decrease rapidly, and the air immediately above is cooled through conduction aided by turbulence. Once the air is cooled below its dew-point temperature, condensation occurs and radiation fog is produced. Since the length of the cooling period is critical, radiation fog is more common during the autumn and winter in mid and high latitudes, e.g. in the British Isles. The cooling of the air by the surface results in a progressive decrease in the environmental lapse rate and the atmosphere eventually becomes stable. A ground level inversion may well develop (Fig. 3.5(2)).

Although conditions favourable for the formation of radiation fog may exist over the sea, the diurnal ranges of both sea surface and air temperatures are too small for the air to be cooled below its dew-point temperature, and therefore no condensation occurs. Radiation fog will affect visibility at sea if it drifts over estuaries and coastal waters as a result of light offshore winds. The air will then be heated by the warmer water over which it passes and, as its temperature increases, the fog droplets will gradually evaporate.

Radiation fog may disperse as a result of an increase in land surface temperature during the day, since the surface heats the air immediately above and lowers its relative humidity. An increase in wind speed can also cause dispersal since it overturns the air. The RH is thus decreased as a result of mixing with unsaturated air. As the increase in land surface temperature is small during the winter day, fog may persist on land and continue to affect estuaries and coastal waters.

In tropical regions, radiation fog is comparatively rare at sea level, but may be experienced over river estuaries during the early hours of the morning. The fog develops during the night over adjacent river banks, where the air has a high relative humidity due to the presence of open water. Similar conditions also apply in swamp and marsh areas.

HAZE

Development of haze on land can often be related to the diurnal cycle of air and surface temperatures. With a significant increase in surface temperature during the day, convection and turbulence increase and dust particles are raised from the surface. Thus the visibility decreases progressively through the day as an increasing number of particles remain suspended in the atmosphere. The condition is more marked when the E.L.R. limits the ascent of air to the lower part of the troposphere, thus causing a rapid increase in the concentration of dust particles. During the night the particles settle onto the surface, and visibility is improved in the early part of the day.

At sea, haze is often the result of movement of dust and sand particles from land. A particular example occurs, usually between November and May, off the north-west coast of Africa, where large quantities of dust from the Sahara are blown off-shore by the local wind, the *Harmattan*. A similar condition occurs in the Gulf during the summer with the *Shamal,* a north-westerly wind which collects dust from the desert. Salt particles, which are potential condensation nuclei, may also produce haze at sea when the RH value of the atmosphere is too low for water droplets to develop.

CHAPTER 7

WIND

DEFINITION

Wind is defined as the horizontal movement of air across the surface of the earth. The direction from which it blows and its speed are its important characteristics. Wind direction is related to True North (360°(T)), and it is recorded and coded in meteorological reports to the nearest 10°(T) clockwise from this direction. A wind blowing from the East will be recorded as 090°(T), whilst no wind or *calm* condition is recorded as 00. Wind speed is currently expressed in knots or metres per second.

The wind is rarely steady in direction and speed for any prolonged period, and thus a mean value over a ten minute period is assessed for observational purposes. A *gust,* an increase, and a *lull,* a decrease in speed about the mean value may well be experienced. If, during the ten minute period, the wind speed changes by 10 knots or more, and the new speed is maintained for more than 3 minutes, then the new speed is recorded. A *squall* is a prolonged gust with a duration of more than one minute and an increase in speed of at least 16 knots, or three steps on the Beaufort Scale (see below), its speed being Force 6 or greater.

OBSERVATION

Land Observations

The standard observing height for wind is 10 metres above the surface. A wind vane, used to indicate direction, and an anemometer for speed are mounted together on a mast where there is a free flow of air across the surface (Fig. 7.1). The operation of the wind vane is dependent upon the pressure of the wind acting on the vertical flat plate (A) so that the bar (B), which rotates horizontally and freely about its centre, is aligned with the direction of the wind. The other end of the bar is the pointer which indicates the direction from which the wind blows. The position of the pointer can be registered on a digital or dial display, and can be continuously recorded on a chart, often sited at some distance from the mast itself.

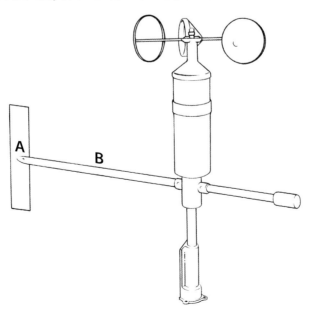

Fig. 7.1 Wind vane and anemometer.

Force	Description	Sea criterion	Wind speed (knots)		Wind waves (sea state)		Plate
			Mean	Limits	Description	Height (m)	
0	Calm	Sea like a mirror.	0	<1	Calm	0−0.1	30
1	Light air	Ripples with appearance of scales are formed, but without foam crests.	2	1−3			31
2	Light breeze	Small wavelets, still short but more pronounced. Crests have a glassy appearance and do not break.	5	4−6	Smooth	0.1−0.5	32
3	Gentle breeze	Large wavelets. Crests begin to break. Foam of glassy appearance. Perhaps scattered white horses.	9	7−10			33
4	Moderate breeze	Small waves, becoming longer; fairly frequent white horses	13	11−16	Slight	0.5−1.25	34
5	Fresh breeze	Moderate waves, taking a more pronounced long form; many white horses are formed. Chance of some spray.	19	17−21	Moderate	1.25−2.5	35
6	Strong breeze	Large waves begin to form; the white foam crests are more extensive everywhere. Probably some spray.	24	22−27	Rough	2.5−4.0	36
7	Near gale	Sea heaps up and white foam from breaking waves begins to be blown in streaks along the direction of the wind.	30	28−33	Very rough	4.0−5.5	37
8	Gale	Moderately high waves of greater length; edges of crests begin to break into spindrift. The foam is blown in well-marked streaks along the direction of the wind.	37	34−40	High	5.5−7.5	38
9	Strong gale	High waves. Dense streaks of foam along the direction of the wind. Crests of waves begin to topple, tumble and roll over. Spray may affect visiblity.	44	41−47	Very high	7.5−11.5	39
10	Storm	Very high waves with long overhanging crests. The resulting foam, in great patches, is blown in dense white streaks along the direction of the wind. On the whole the surface of the sea takes on a white appearance. The "tumbling" of the sea becomes heavy and shock-like. Visibility affected.	52	48−55			40
11	Violent storm	Exceptionally high waves (small and medium-sized ships might be for a time lost to view behind waves). The sea is completely covered with long white patches of foam lying along the direction of the wind. Everywhere the edges of the wave crests are blown into froth. Visibility affected.	60	56−63	Phenomenal	>11.5	41
12	Hurricane	The air is filled with foam and spray. Sea completely white with driving spray; visibility very seriously affected.	—	>64			42

Note: Wave height data are for open sea conditions. In enclosed waters or when near land with an offshore wind, wave heights will be smaller and the waves steeper.

Table 7.1 The Beaufort Scale.

The anemometer consists of three hemispherical cups with beaded edges held vertically on horizontal bars of equal length, which radiate at 120° apart from a central pivot (Fig. 7.1). The cups are designed to achieve a turbulent-free flow of air. As the pressure of the wind on the concave side of the cups is greater than on the convex side, the cups rotate and operate a small electrical generator which produces a current. The value of the current is related to the speed of rotation and hence the wind speed. Thus the wind speed can be registered in the same way as the wind vane data.

Sea Observations

In the 19th Century, Admiral Beaufort introduced the *Beaufort Scale,* a scale from 0–12 called *Beaufort Force* numbers, each number corresponding to a range of wind speeds. Originally expressed in terms of the effect on a man of war and the setting of sails, today it is based on the state of the sea (Table 7.1 and Plates 30–42)). The sea criteria are the wind waves which are generated by the wind which has been in existence for a reasonable period, and having an adequate *fetch* (the distance of open water over which the wind has blown). However, the wind is not the only factor influencing the sea state, and allowances should be made for tides, currents, depth of water and precipitation, where these are seen to affect the sea state. Tides opposing the direction of the wind waves will create more "lop", and an overestimate of wind speed is possible. Heavy precipitation flattens the sea and may lead to an underestimation. Wind direction is established by observing the direction from which the wind waves advance.

Waves generated by winds at some distance from, or at some time previously at the point of observation also affect the sea. This wave motion is called *swell,* and is excluded when recording the Beaufort Force. Swell, in contrast to wind waves, has a long and generally low regular wave form (Plate 43). It may be at any angle to the wind waves, and more than one swell may exist at the same time. Thus swell gives a useful indication of conditions existing, or which existed at some distance in the direction from which it is coming. Swell generated by wind conditions in the Southern Oceans has often been observed in the Western Approaches to the British Isles. The presence of swell may also be one of the earliest indications of a tropical cyclone.

For meteorological purposes the period and height of wind waves and swell, and the direction from which the swell is coming are recorded. The period of a wave is the time taken for the passage of two successive crests past a point selected by the observer. The height of the wave is the vertical distance between the bottom of the trough and the top of the crest. As waves normally occur in groups, the height and period are assessed by observing two or more of the relatively large waves in each group, until at least ten such waves have been observed. The period is the average value of the recorded times, and the height is the average value of the heights observed. Table 7.1 includes a guide to the classification of wind waves related to wind speed. The following table is a guide to the classification of swell waves:

Length		Height	
Description	Length (m)	Description	Height (m)
Short	0–100	Low	0–2
Average	100–200	Moderate	2–4
Long	>200	Heavy	>4

Table 7.2 Swell Waves.

Conditions at sea often preclude an accurate observation of the sea state and the use of the Beaufort Scale. On these occasions the wind vane and anemometer, funnel smoke, and flags can be observed to establish the relative wind speed and direction. As the true wind is required, it is derived using a vector triangle (Fig. 7.2), where the reversal of the vessel's course of 290°T, represents the vessel's wind direction which is plotted with the relative wind of 045°T, using the vessel as the focal point of the vector triangle. The length of each line is scaled to represent speed. The third side of the triangle gives the true wind direction and speed.

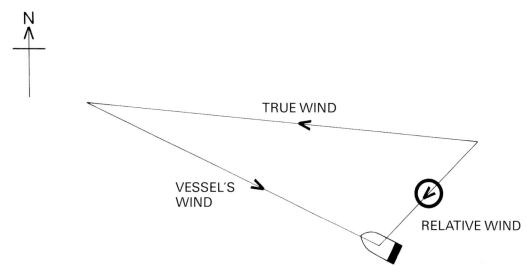

Fig. 7.2 True wind vector triangle.

If there is no relative wind, the vessel's course and the speed reversed is the same as the true wind direction and speed (Relative Wind – 00: Course 090°(T) – 10 knots: True Wind 270°(T) – 10 knots).

If the relative wind is from the bearing on which the ship is steering, the direction of the true wind is the vessel's course, and its speed is the difference between the vessel's speed and the relative wind speed (e.g. Relative Wind 270°(T) – 20 knots: Course 270°(T) – 10 knots: True Wind 270°(T) – 10 knots).

The siting of wind vanes and anemometers on a vessel pose problems in providing a free flow of air. The standard height of 10 metres is rarely practicable if the effect of the superstructure is to be avoided, and a compromise has to be accepted, with the consequent inaccuracy of readings on some bearings.

LARGE SCALE AIR FLOWS

Surface winds may be either large scale air flows associated with pressure systems (Chapter 8), or air flows associated with local conditions, or a combination of both. The observation and recording of atmospheric pressure enable the distribution of pressure over an area of the surface to be established (Chapter 2).

Pressure Gradient

The change of pressure over unit distance at right angles to the isobars is termed the *horizontal pressure gradient* (Fig. 7.3). The gradient is *steep* when the isobars are close together, and *slack* when they are far apart, both terms being used in a relative sense and without absolute values.

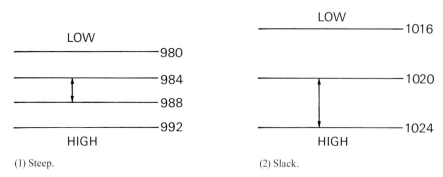

Fig. 7.3 Horizontal pressure gradient.

Pressure Gradient Force

When a horizontal pressure gradient exists, a force, termed the *pressure gradient force,* acts on the air which moves from high to low pressure at right angles to the isobars (Fig. 7.4). However, for a number of reasons the actual air motion observed at the surface is seldom in the direction of, or at the speed related to, this force.

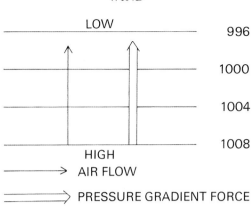

Fig. 7.4 Pressure gradient force and air flow.

Coriolis Force

The Coriolis force or the geostrophic force, so termed because it relates to the rotation of the earth about its axis, causes an air particle to be deflected to the *right* of its line of motion in the *northern hemisphere,* and to the *left* of its line of motion in the *southern hemisphere.* The force always acts at right angles to the line of motion of a particle. Mathematically it can be shown that in the horizontal plane, with the xy coordinates aligned to north and east respectively, the Coriolis force per unit mass is $2\Omega \sin \Phi v$ where Ω is the angular velocity of the earth, Φ the latitude, and v the speed of the air. It follows that at the equator $\sin \Phi = 0$ and the force equals zero. In low latitudes the rate of change of the value of the force will be at a maximum for a given wind speed. As latitude increases the value of the force increases, but its rate of change decreases.

Geostrophic Wind

When isobars are straight lines parallel to each other, the resultant horizontal motion, due to the action of the pressure gradient and Coriolis forces, is termed the *geostrophic wind* (Fig. 7.5). Its direction is parallel to the isobars and its speed is constant. The geostrophic wind is used by professional meteorologists in forecasting, but as it does not exist at the surface, it is of limited practical value to the seafarer.

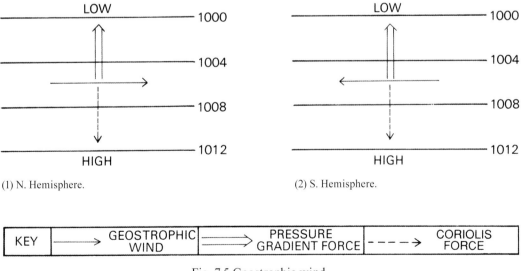

Fig. 7.5 Geostrophic wind.

Cyclostrophic Force

When an air particle is following a curved path it will be subject to the *cyclostrophic force* which acts radially outwards from the centre of rotation. The value of the force is directly dependent upon the speed of the air particle.

Gradient Wind

The *gradient wind* is the horizontal air motion parallel to isobars which are curved. Air movement initiated through the action of the pressure gradient force is subject to a deflection due to the Coriolis force. Due to the curved isobars the flow will begin to follow a curved path around the pressure system hence it will also be subject to the cyclostrophic force (Fig. 7.6). The resulting direction of the gradient wind in each hemisphere is as follows in Table 7.3. Assuming fixed values for the pressure gradient, Coriolis and cyclostrophic forces, it follows that the gradient wind speed will have a larger value in an area of high pressure compared to that in an area of low pressure.

Pressure System	N Hemisphere	S Hemisphere
Low	Anticlockwise	Clockwise
High	Clockwise	Anticlockwise

Table 7.3 Gradient wind.

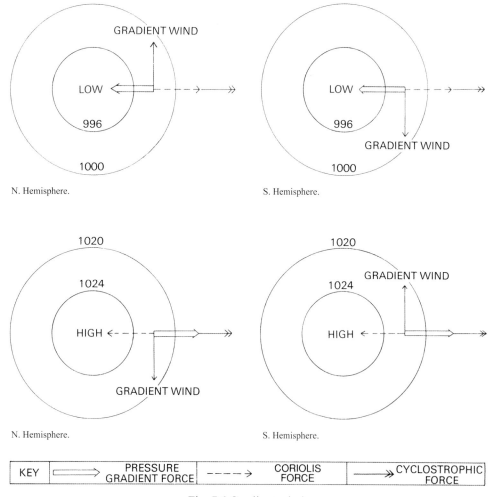

Fig. 7.6 Gradient wind.

Friction

Air moving across the surface of the earth is affected by *friction,* and does not achieve the speed which in theory is directly related to the horizontal pressure gradient. As a result, the Coriolis and cyclostrophic forces have smaller values, and therefore neither geostrophic nor gradient winds exist. The pressure gradient force becomes dominant, and the net result is a cross-isobaric component of the surface air flow from high to low pressure which is the surface wind. The angle between the surface wind and the isobar is termed the *angle of indraught* which is 10° to 15° over the sea (Fig. 7.7). Over land the effect of friction is greater, and the angle of indraught is therefore larger.

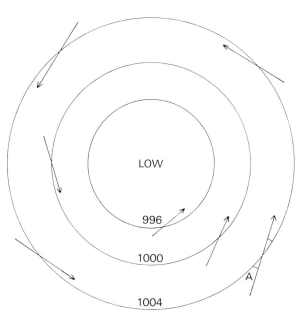

A – ANGLE OF INDRAUGHT

Fig. 7.7 Angle of indraught.

Buys Ballot's Law

In 1857, Buys Ballot formulated a law identifying the relationship between wind and pressure distribution. The law states that if an observer has his back to the wind, then low pressure will be to the *left* in the *northern hemisphere* and to the *right* in the *southern hemisphere*. It follows that high pressure will lie on the right in the northern and on the left in the southern hemisphere (Fig. 7.8(1)). An alternative way in which the law may be expressed is if an observer faces the direction from which the wind is blowing. Low pressure will lie to the right in the northern hemisphere and to the left in the southern hemisphere (Fig 7.8(2)). Under certain circumstances the number of points of the compass to the right or left of the wind in which the low pressure centre lies is important, particularly when assessing the position and movement of tropical cyclones (Chapter 9).

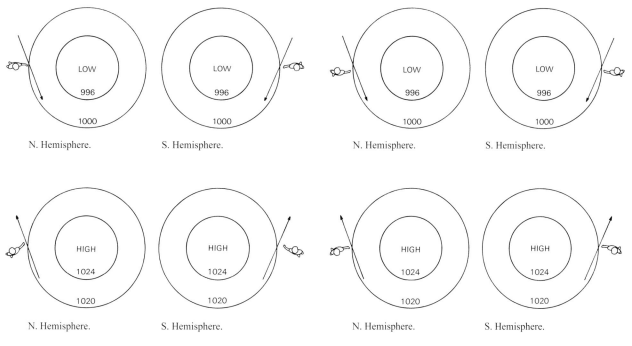

(1) Buys Ballot's Law – Back to the wind.   (2) Buys Ballot's Law – Facing the wind.

Fig. 7.8 Buys Ballot's Law.

Surface Wind

Although a geostrophic wind does not exist at the surface, the concept is used in determining surface wind speeds. Facsimile meteorological charts may have geostrophic wind scales specifically related to the projection and scale of each chart. Fig. 7.9 is an example of the use of the scale in determining surface wind speed at 48.5°N. The perpendicular distance between isobars (AB) is measured and plotted on the scale. The geostrophic wind speed of 12 knots (C), found by interpolation, is reduced by one third to allow for friction, and the predicted surface wind speed is thus 8 knots.

In Fig 7.9(2)), an observer will be facing the direction from which the wind is blowing when lower pressure values are to his right. With the isobars lying approximately WSW- ENE and an angle of indraught of 10° to 15° the wind direction will be south-west.

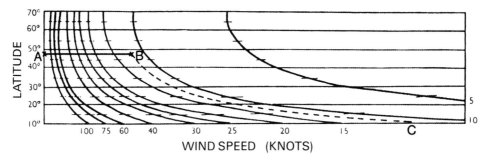

(1) Geostrophic wind scale for 4 hPa isobaric interval.

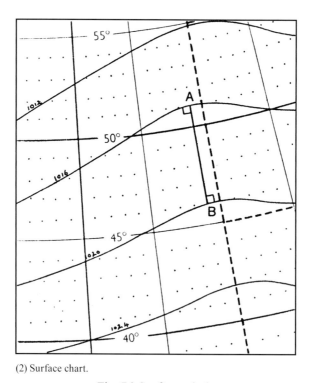

(2) Surface chart.

Fig. 7.9 Surface wind.

SEA AND LAND BREEZES

The seafarer frequently encounters local wind conditions in coastal waters resulting from heating and cooling of adjacent land and sea surfaces. Such conditions also occur over comparatively large areas of water inland (e.g. The Great Lakes).

During the day, the land surface temperature increases more rapidly than that of the sea surface (Chapter 3). Air temperature above the land is therefore higher, and the vertical motion of the air through convection modifies the pressure gradient in the horizontal plane (1–2) (Fig. 7.10). As the pressure above the land (1) becomes greater than that over the sea in this plane, so the air moves towards lower pressure (2).

The surface air pressure on land (4) then decreases, and air moves from the sea (3) to land (4). Thus during the day a *sea breeze* blows onshore, which may be felt for some distance inland, and an offshore wind blows at some 1000 metres above the sea. The sea breeze gradually increases in speed as the land surface temperature increases, and reaches a maximum by mid-afternoon. When the sky is clear a large amount of solar radiation reaches the surface and the wind speed will be at its maximum (mid-latitudes – Force 3; low latitudes – Force 5).

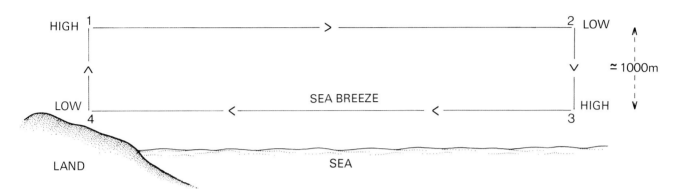

Fig. 7.10 Sea breeze.

The sea breeze has a considerable influence on weather conditions at the coast and for some distance inland. As the sea breeze is relatively cool, coastal zones experience lower air temperatures compared with those inland. The relative humidity value at the coast is higher than inland, and increases during the afternoon when the breeze reaches its maximum speed. This condition contrasts with diurnal variation of relative humidity noted in Chapter 4. The leading onshore edge of the sea breeze is often marked by a line of cumuliform (convective) cloud, which may result in precipitation, particularly in tropical areas.

By contrast the *land breeze* which develops during the night is slight, reaching Force 1 – 2. With clear skies the radiative cooling of the land surface is at a maximum, but the sea surface temperature decreases slowly (Chapter 3). Above the land the air cools, becomes denser (5), and flows downslope towards the sea (6), hence creating the land breeze (Fig. 7.11). The air, heated by the warmer sea surface, ascends (7) and then moves towards the shore (8). Land breezes are normally evident by midnight local time, but on occasions may not develop until the early hours of the morning. It should be noted that the small temperature difference at night between land and sea surfaces, and the topography of the land, both influence the speed of the land breeze. In tropical areas the convective process (6 → 7) is marked by the development of cumuliform cloud over the sea.

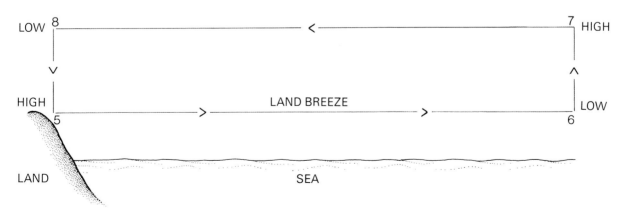

Fig. 7.11 Land breeze.

It is worth noting that both land and sea breezes are caused by local pressure gradients and flow down the gradient. In mid-latitudes the wind may eventually orientate itself parallel to the coastline, reflecting the influence of the Coriolis force on a well developed local system.

KATABATIC AND ANABATIC WINDS

Katabatic (downslope) winds affect sea conditions off mountainous coastal areas, particularly the Norwegian fjords and the ice covered regions of Greenland and Antarctica. They develop at night when there are clear skies, a general slack pressure gradient, and rapid radiative cooling of the land. The air at X (Fig. 7.12) adjacent to the slope, becoming cooler and therefore denser than the air which is further away at the same level (Y), descends forming the katabatic wind. On reaching the foot of the slope the wind moves out to sea. Adiabatic warming of the air during descent is counteracted by conduction as it is in continuous contact with the colder mountain slope. When the slope is covered with ice or snow, which are effective insulators, very limited conduction takes place between the slope and the upper surface of the ice or snow cover. Thus during the cooling period overnight the upper surface experiences a more rapid decrease in temperature than that of a bare slope. Under such conditions, the adjacent air, becoming very cold and dense, descends at a speed which can reach gale force conditions.

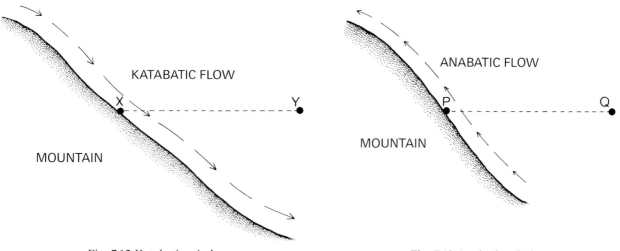

Fig. 7.12 Katabatic wind.   Fig. 7.13 Anabatic wind.

During the day an *anabatic* wind develops. With clear skies the slope absorbs solar radiation and heats the air (P) directly in contact. This air then flows up the slope, as it is at a higher temperature than the air (Q) further away at the same level (Fig. 7.13). The air moving up the slope is subject to expansion and therefore cooling. The contact maintained with the warm slope ensures its continued ascent, with a maximum speed during mid-afternoon. Anabatic winds occur in the Alps in summer months when there is generally a slack pressure gradient.

The need for seafarers to understand land and sea breezes is clear. Sea breezes can reach significant speeds which, combined with current and tide may create hazardous conditions. Sailing directions should always be consulted.

COLOUR PLATES

METEOROLOGICAL PHENOMENA & SEA STATES

A – CLOUDS AND METEOROLOGICAL PHENOMENA

B – SEA STATES AND STORM WAVES

Plate 1 Cirrus (Ci).

Plate 2.1 Cirrostratus (Cs).

Plate 2.2 Cirrostratus (Cs) with halo.

Plate 3 Cirrocumulus (Cc).

Plate 4 Altostratus (As).

Plate 5 Altocumulus (Ac).

Plate 6 Stratus (St).

Plate 7 Stratocumulus (Sc).

Plate 8 Nimbostratus (Ns).

Plate 9 Cumulus (Cu).

Plate 10 Cumulus (Cu).

Plate 11 Cumulonimbus (Cb).

Plate 12 Cumulonimbus – developing from cumulus.

Plate 13 Cumulonimbus – developing from cumulus after ten minutes.

Plate 14 Cumulonimbus – developing from cumulus after forty minutes.

Plate 15 Cumulus (Cu) and Stratocumulus (Sc).

Plate 16 Waterspout.

Plate 17 Orographic cloud.

58

Plate 18 Lenticular cloud.

Plate 19 Cumulonimbus precipitating.

Plate 20 Lightning.

Plate 21 Entering sea fog in the English Channel – haze but no fog.

Plate 22 Entering sea fog in the English Channel – fog developing.

Plate 23 Entering sea fog in the English Channel – fog.

Plate 24 Arctic Sea Smoke – St Lawrence Seaway.

Plate 25 Pampero.

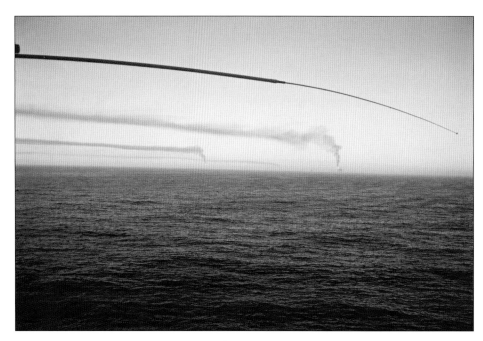

Plate 26 Effect of a temperature inversion – North Sea.

Plate 27 Trade wind cumulus.

Plate 28 Rainbows – primary and secondary.

62

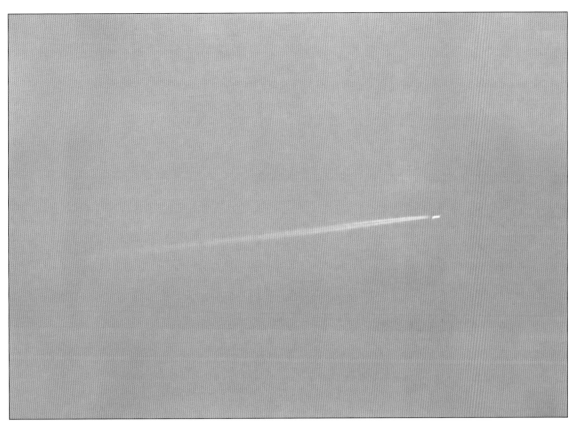

Plate 29.1 Contrails.

Plate 29.2 Contrails.

Plate 30 Force 0.

Plate 31 Force 1.

Plate 32 Force 2.

64

Plate 33 Force 3.

Plate 34 Force 4.

Plate 35 Force 5.

65

Plate 36 Force 6.

Plate 37 Force 7.

Plate 38 Force 8.

66

Plate 39 Force 9

Plate 40 Force 10.

Plate 41 Force 11.

Plate 42 Force 12.

Plate 43 Swell and wind waves Force 2–3.

Plate 44 Extreme storm wave –North Atlantic, Force 12.

CHAPTER 8

TEMPERATE AND POLAR ZONE CIRCULATION

GENERAL CIRCULATION OF THE ATMOSPHERE

The general circulation of the atmosphere is the three dimensional flow of air on a global scale, with an associated transfer of energy. It is a highly complex system and subject to change both in space and time.

If it is assumed that the surface of the earth is uniform and the time of year equinoctial (spring or autumn), the idealized distribution of surface pressure and associated air flow will be as shown in Fig. 8.1.

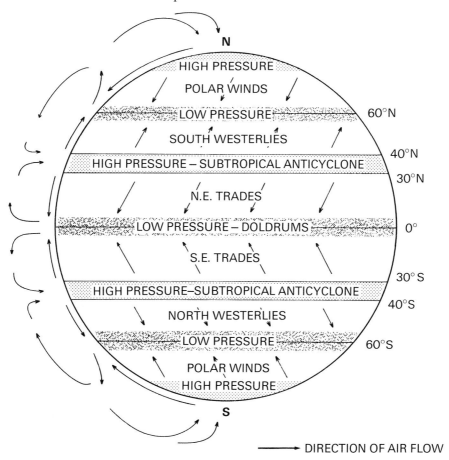

Fig. 8.1 General circulation of the atmosphere – idealized.

There is convergence, a net accumulation of air, at the surface at 0° and 60°N and S where the pressure is low. However there is divergence, a net loss of air, at the surface between 30° and 40°N and S, and at the poles where the pressure is high. In general, where there is convergence at the surface, there is divergence in the upper troposphere and air ascends between the two levels. Conversely air descends where there is divergence at the surface, and convergence in the upper troposphere.

At 60°N and S two masses of air having different temperatures converge. The boundary between these masses is called a *frontal zone or frontal surface,* and where it intersects the surface of the earth it is called a *front* (Fig. 8.2). The frontal zone is inclined at an angle to the surface with the warm air tending to rise over the cold air.

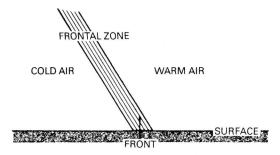

Fig. 8.2 Frontal zone – vertical section.

The transfer of energy by the general circulation is related to the net radiation balance of the earth/atmosphere system, the balance being the difference between the total incoming solar radiation and outgoing terrestrial radiation. Annually there is a surplus of energy between 40°N and 40°S and a deficit in higher latitudes. However, the amount of energy transferred varies with the time of year.

Some of the major features of the general circulation can be identified on surface synoptic charts. Fig. 8.3 is a typical example for the North Atlantic area encompassing the temperate (40°N–66°N), parts of the subtropical (23°N–40°N), and the polar zones (66°N–90°N). While the subdivision into zones may assist in the analysis of atmospheric features, it must not be forgotten that it is a manmade convention.

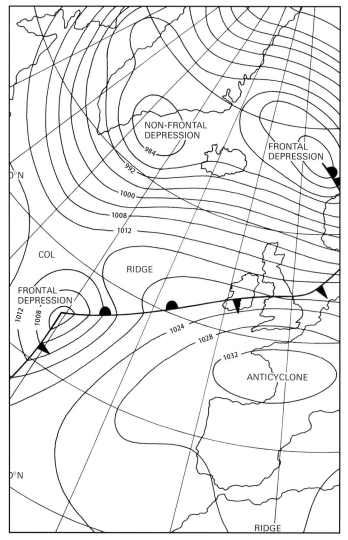

Fig. 8.3 Temperate and polar zone circulation – surface synoptic chart.

Pressure systems in temperate and polar zones play a critical part in determining weather and sea state conditions, and an understanding of the significance of the features on synoptic and prognostic (forecast) charts and in weather bulletins is essential.

FRONTAL DEPRESSIONS

A frontal depression is characterized by having a surface pressure distribution marked by one or more isobars enclosing an area of low pressure. The value of the pressure at the centre, which varies throughout the life cycle of the depression and from one depression to another, generally ranges between 950 hPa and 1020 hPa. A depression having several isobars and a central pressure value below 950 hPa is termed a *deep depression.*

Most frontal depressions develop on a frontal zone (Fig. 8.4). *Polar fronts* are boundaries between polar and tropical air; *arctic fronts* between arctic and polar air; *antarctic fronts* between antarctic and polar air; the *Mediterranean front* between polar and tropical air. The mean positions of these frontal zones are shown in Fig. 8.4 for January and July. In January the polar fronts in the southern hemisphere are aligned approximately east-west. In contrast the polar fronts in the northern hemisphere are aligned with the eastern coastlines of North America and Asia, partly reflecting the contrasting influence of land and sea. In the Mediterranean the contrast in air masses is apparent in January, but scarcely exists in July when there is no frontal zone. In July frontal zones in the northern and southern hemispheres lie approximately east-west.

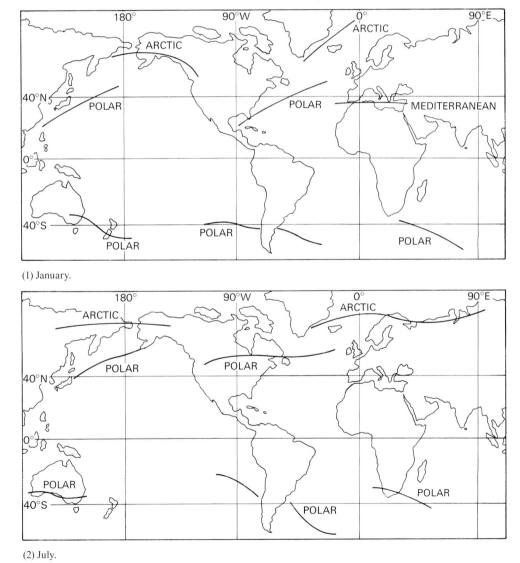

(1) January.

(2) July.

Fig. 8.4 Mean position of frontal zones.

Life Cycle of a Frontal Depression

The formation, development and decay of a frontal depression, whose life cycle may vary in length from three to seven days, can be illustrated by a series of plan diagrams of the system at the surface of the earth (Fig. 8.5(1) to (6)).

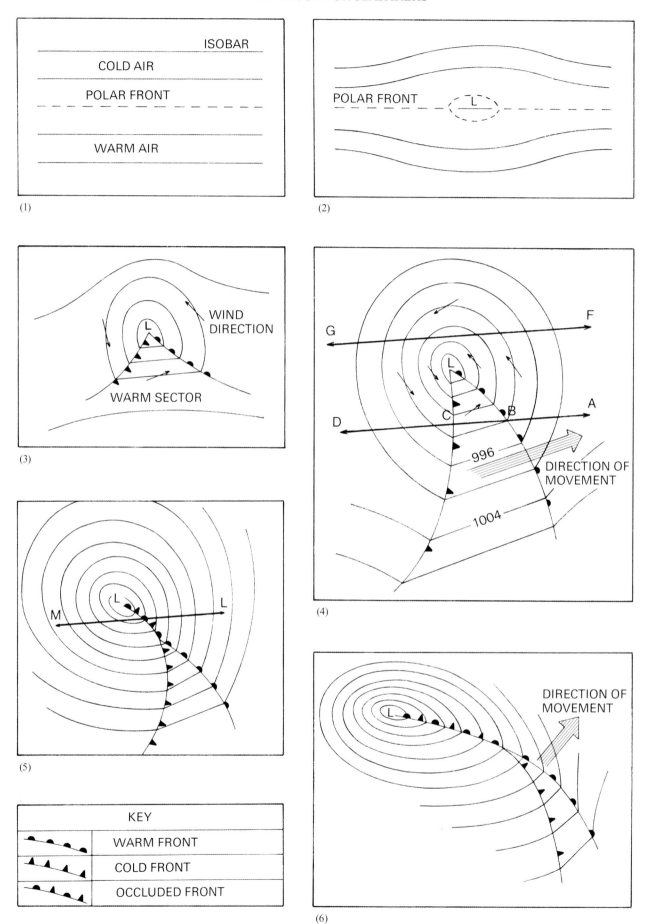

Fig. 8.5 Life cycle of a polar front depression, surface plan view.

The progressive separation of isobars indicates a decrease in pressure at a point on a frontal zone (Fig. 8.5(1) and (2)). A wave appears on the front and, if unstable, it continues to develop. The centre of low pressure is at the apex of the wave, and the surface air circulation develops a typical anticlockwise (northern hemisphere) and inward movement, and *frontogenesis* commences reflected by the steepening of the horizontal temperature gradients across the front. Sections of the original front are now identified as the *warm* or *cold* front with an intervening warm sector (Fig. 8.5(3)). During these initial stages the amplitude of the wave increases, and the depression moves rapidly eastwards parallel to the isobars of the warm sector, sometimes at speeds of over 40 knots (Fig. 8.5(3) and (4)). The position of the warm front at the surface marks the point at which warm air replaces cold air; the position of the cold front where cold air replaces warm air.

Within 24 hours of the initial wave formation, the depression usually reaches maturity (Fig. 8.5(4)) and its central pressure value has decreased (the depression has *deepened*). As the depression deepens so the pressure gradient becomes steeper and the wind speeds increase.

The speed of advance of the cold front of the depression is greater than that of the warm front; hence the area of the intervening warm sector diminishes. As the warm sector is progressively lifted off the surface from the centre of the depression outwards an *occluded front,* also termed an *occlusion,* develops (Fig. 8.5(5) and (6)). The occluded front may be described as being warm or cold as explained later in Chapter 8. Into the early stages of occlusion the frontal depression will continue to deepen. Subsequently its central pressure value will increase (the depression is *filling*), and simultaneously wind speeds decrease as the pressure gradient slackens. The speed of advance of the depression decreases and eventually it may become stationary with the occluded front pivoting around its centre (Fig. 8.5(6)). Eventually the weak low pressure area and occluded front fade. The process from initial occlusion to the final fading of the front is termed *frontolysis.*

The life cycle of a frontal depression can be observed by studying a series of synoptic or prognostic surface charts. In Fig. 8.6(1)-(3), covering a period of 48 hours, depression A develops on the polar front and moves NNE, deepening as the amplitude of the wave increases.

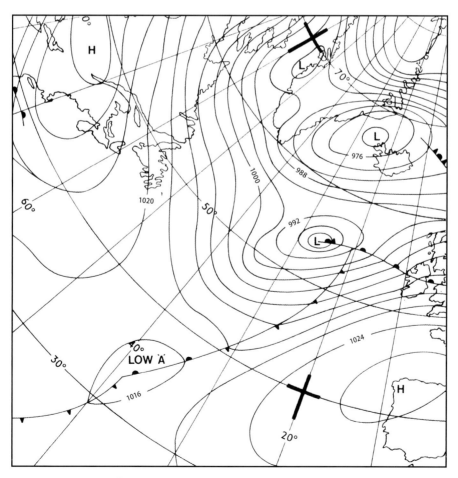

Fig. 8.6(1) Synoptic chart – 1200 UTC, Day 1.

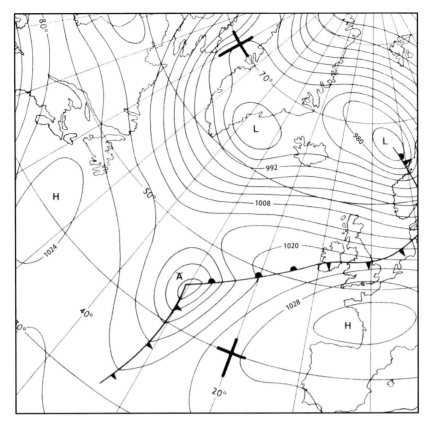

Fig. 8.6(2) Synoptic chart – 1200 UTC, Day 2.

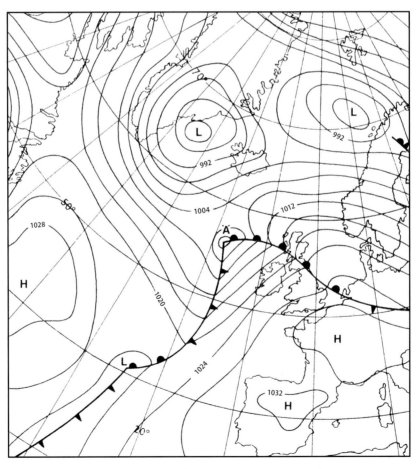

Fig. 8.6(3) Synoptic chart – 1200 UTC, Day 3.

Jet Stream

If a depression is to develop, the air flow in the upper troposphere must be divergent. This divergent flow is associated with a *jet stream,* a fast moving current of air with strong lateral and vertical wind shears which is located immediately below the tropopause. Fig. 8.7 shows the relationship between the polar front jet stream and a frontal depression during the life cycle of the latter.

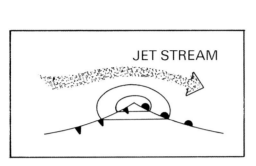

(1) Early stages of development of depression – plan view.

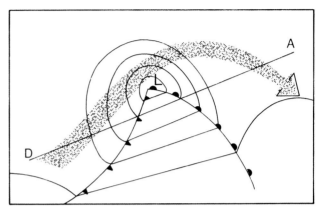

(2) Depression fully developed – plan view.

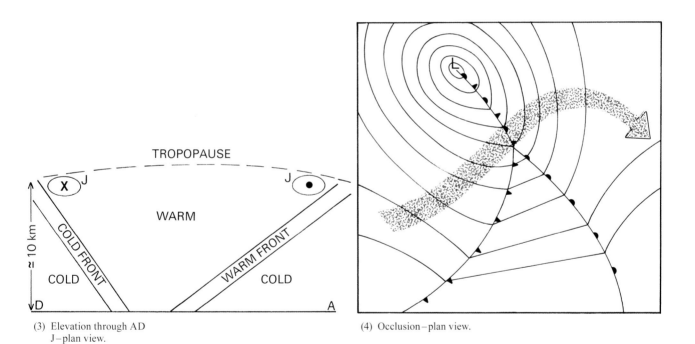

(3) Elevation through AD J – plan view.

(4) Occlusion – plan view.

Fig. 8.7 Polar front jet stream and a frontal depression.

Anafronts and Katafronts

Within a frontal depression the warm air usually ascends relative to cold air. This movement is marked by the presence of active warm and cold fronts termed *anafronts.* The clouds which develop are shown in the elevation in Fig. 8.8(2) corresponding to line AD in Fig. 8.8(1). On occasions the warm air may descend relative to the cold air, and a relatively inactive front termed a *katafront* exists, with limited low level stratiform cloud.

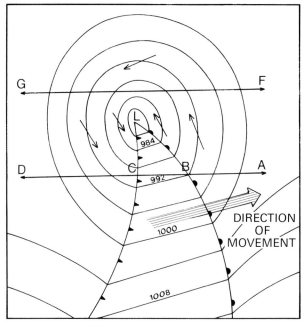

(1) Plan of frontal depression.

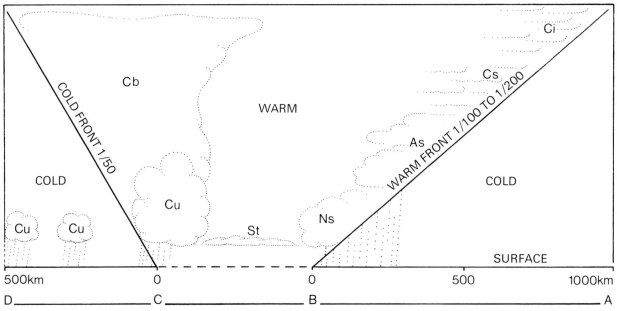

(2) Elevation through AD in (1) above.

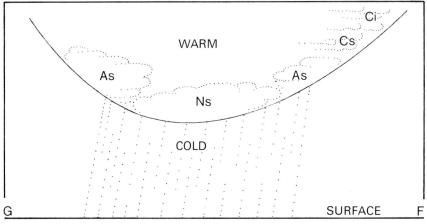

(3) Elevation through FG in (1) above.

Fig. 8.8 Plan and elevation of a typical frontal depression in the N. Hemisphere.

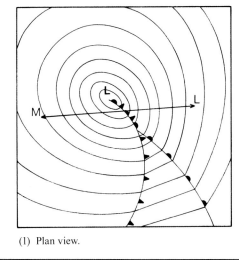

(1) Plan view.

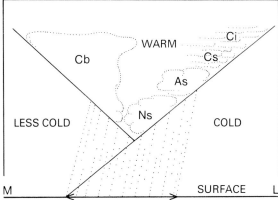

(2) Warm occlusion.

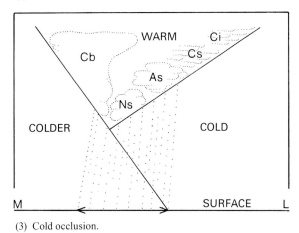

(3) Cold occlusion.

Fig. 8.9 Warm and cold occlusions.

Warm and Cold Occlusions

Occlusions may be either warm or cold depending upon the temperature difference between the cold air originally behind the cold front and that of the cold air ahead of the warm front. At the beginning of the life cycle of a depression the two masses of cold air may well have been at the same temperature. However, the subsequent development and movement of a depression incorporates cold air from other sources resulting in the temperature difference.

If the air behind the occluded front is less cold than the air ahead, the occlusion is a *warm* occlusion (Fig. 8.9(2)). If the air is colder behind the front, then it is a *cold* occlusion (Fig. 8.9(3)). The cloud types associated with recently developed warm and cold occlusions are the same as those of the warm and cold fronts. With older occlusions, cloud and precipitation will be less significant as a result of frontolysis.

Weather Conditions

The wind, air temperature and precipitation associated with frontal depressions are of particular concern to the seafarer. The sequence of weather will depend on the position of the observer relative to the centre of the depression, and the following tables are a guide:

Element	Sequence
Pressure	Decreases then increases
Wind	Backs and increases then decreases
Temperature	Steady, but may decrease when wind backs to N and NW
Cloud	Ci, Cs, As, Ns, As
Weather	Fine, followed by rain increasing, then decreasing
Visibility	Good, deteriorating then improving

Table 8.1 Weather sequence of a frontal depression to the south
of the observer in the N. Hemisphere (Fig. 8.8(1) and (3)).

Element	Sequence				
	In advance of warm front	At warm front	In warm sector	At cold front	To rear of cold front
Pressure	Decreases	Stops decreasing	Little change	Rapid increase	Increases
Wind	Increasing may back	Veers and sometimes decreases	Steady	Veers, squally	May veer further and decreases
Temperature	Increases	Increases	Steady	Decreases	Steady, may decrease when wind veers further
Cloud	Ci, Cs, As, Ns	Ns	St	Cu,Cb	Cu,Cb
Weather	Fine, then rain increasing	Rain almost or completely stops	Cloudy, drizzle	Hail, rain, lightning, thunder	Showers
Visibility	Fine, then deteriorating	Poor, frontal fog	Poor, advection fog	Poor	Improves, good except in showers

Table 8.2 Weather sequence of a frontal depression to the north
of the observer in the N. Hemisphere (Fig. 8.8(1) and (2)).

The following comments amplify Tables 8.1 and 8.2:

(a) If the course of the observer is westwards then the duration of events will be shorter.

(b) Absolute values of pressure, wind speed and temperature, are not given as these will depend upon the particular frontal depression.

(c) Wind direction:

 (i) *Backs* or backing – the wind direction changes anticlockwise e.g. East through North to West.

 (ii) *Veers* or veering – the wind direction changes clockwise e.g. East through South to West.

(d) Rain. The term used includes rain, sleet or snow.

(e) The development of frontal fog is due to the evaporation of rain as it passes through the cold air beneath the frontal surface. As a result the cold air becomes saturated and condensation occurs.

The sequence of weather for cold and warm occlusions in the northern hemisphere is given below:

Element		Sequence		
		Ahead of occluded front	*At occluded front*	*To rear of occluded front*
Pressure		Decreases	Stops decreasing	Increases
Wind		Backs	Veers	May continue to back; decreases
Temperature	Cold	Decrease	Decrease	Steady
	Warm	Increase	Increase	Steady
Cloud		Ci, Cs, As, Ns	Ns, Cb	Cb clearing
Weather	Cold	Precipitation near front	Heavy precipitation	Heavy precipitation then decreases
	Warm	Fine, then precipitation becoming heavy	Precipitation decreasing	Precipitation ceases
Visibility		Deteriorates	Generally poor	Improves

Table 8.3 Weather sequence of cold and warm occlusions
in the N. Hemisphere (Fig. 8.9).

A number of local winds in the Mediterranean are associated with the passage of depressions eastwards (Fig. 8.10). The *Scirocco,* seldom noticeable in the winter, is a warm southerly wind from the Sahara desert. Initially hot and dry, it may become humid after a sea passage and low level stratus cloud develops. If its speed is high it transports dust, resulting in low visibility and subsequently red rain. Over Egypt the wind is known as the *Khamsin.*

The *Vendavales* occurs in the Western Mediterranean, particularly in late autumn and early spring. It is a strong south westerly wind with associated thunderstorms. In contrast the *Levanter* is an easterly wind occurring in the same area between June and October, with orographic cloud developing over the Rock of Gibraltar.

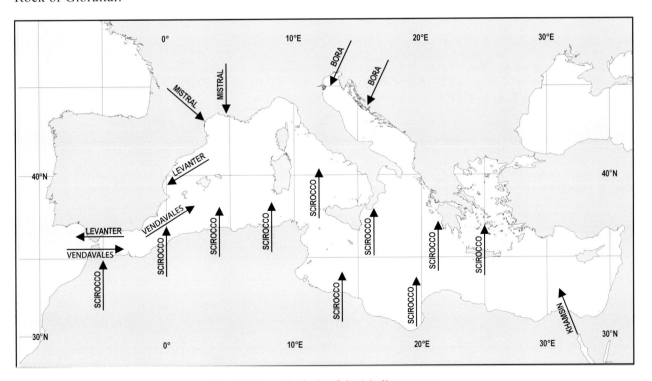

Fig. 8.10 Local winds of the Mediterranean.

The *Mistral,* a cold north to north-westerly wind, occurs in the Gulf of Lions mainly during the winter. It develops as a result of the transit of a depression eastwards over the sea with high pressure over North-West France. The *Bora,* a cold dry north-easterly wind, occurs in the Adriatic during the winter months, and is associated with high pressure over Central Europe and depressions to the south.

Frontal Depressions in the Southern Hemisphere

Polar and antarctic fronts are features of the southern hemisphere circulation and frontal depressions are common (Fig. 8.11). Fig. 8.12 is a surface plan of a depression in the southern hemisphere. Lines AD and FG are directly related to the vertical sections shown in Fig. 8.8(2) and (3) respectively.

The sequence of weather conditions associated with these depressions is similar to those experienced in the northern hemisphere, except for the change in wind direction. In Tables 8.1, 8.2, and 8.3 where the wind is said to *back* in the *northern* hemisphere, it *veers* in the *southern* hemisphere, and vice versa.

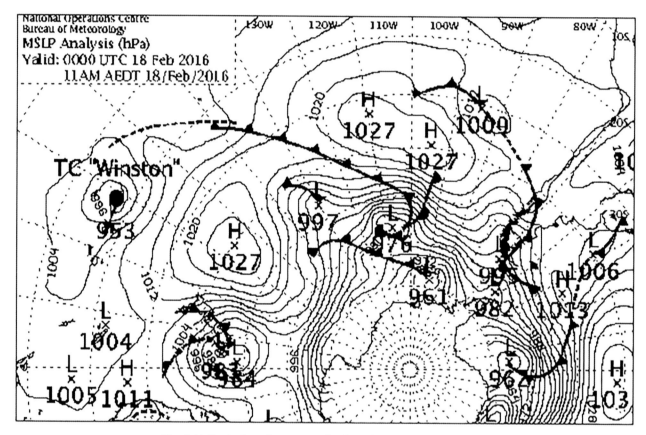

Fig. 8.11 Typical surface synoptic chart for the S. Hemisphere.

Off Southern Australia and South America, the passage of a cold front causes significant changes in weather conditions – a marked decrease in temperature, increased wind speeds, often lightning, thunder and hail, and an accompanying line of low black cloud or *roll cloud.* This is a *line squall,* also known by the local names *Southerly Buster* (Australia) and *Pampero* (South America) (Plate 25).

Family of Depressions

Synoptic charts often show frontal depressions developing in succession along a frontal zone (Fig. 8.6), forming a group known as *a family of depressions.* The number of members in a family varies from four to seven. Each depression forms on the trailing cold front of the preceding depression, which is in the latter stages of its life cycle. Each follows a similar life cycle, but having a path in a lower latitude parallel to that of its predecessor. The end of the family is marked by the invasion of cold air in the form of an anticyclone into lower latitudes. A new family develops when the frontal zone is re-established.

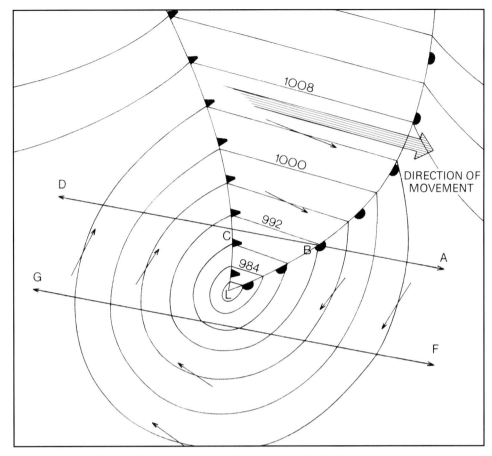

Fig. 8.12 Plan of a frontal depression in the S. Hemisphere.

TROUGHS OF LOW PRESSURE

Troughs of low pressure, which may be frontal or non-frontal, are identified by the form of the isobars extending outwards along an axis from a centre of low pressure. Frontal troughs exist in a frontal depression and are characterized by the general angular form of the isobars (Fig. 8.13(1)) along the line of a marked warm, cold or occluded front (Fig. 8.3).

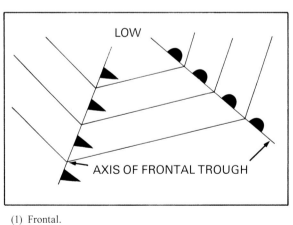

(1) Frontal.

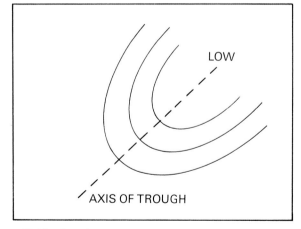

(2) Non-frontal.

Fig. 8.13 Troughs of low pressure.

Non-frontal troughs have more rounded isobaric forms, (Fig. 8.13(2)) and normally develop in the unstable air stream behind a frontal depression (Fig. 8.6(1)). The axis of the non-frontal trough represents an area with a steep pressure gradient and as it advances, an observer in the northern hemisphere can expect the wind to increase in speed and to veer at the axis. A further indication of its passage is increased cumuliform cloud cover and precipitation.

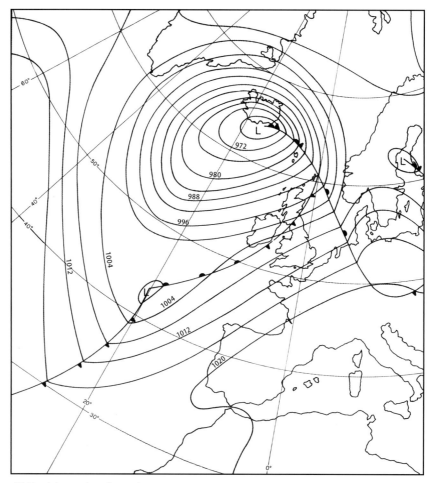

(1) Frontal secondary depression.

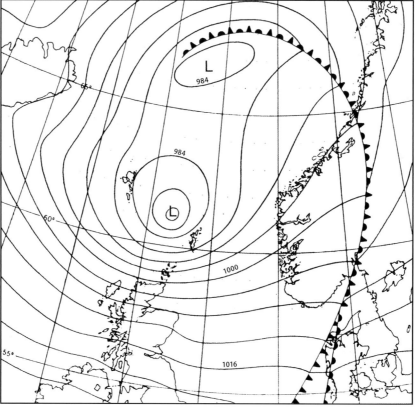

(2) Non-frontal secondary depression.

Fig. 8.14 Secondary depressions.

SECONDARY DEPRESSIONS

A *secondary* depression is one which develops and is embedded within the circulation of a larger depression termed the *primary* depression. The secondary depression moves in the direction of the general air flow around the primary; this movement is termed *cyclonic*. Thus it will move *anticlockwise* in the *northern* hemisphere and *clockwise* in the *southern* hemisphere. The central pressure value of the secondary depression is initially greater than that of the primary. However, as the primary has begun to decay when the secondary begins to develop, the latter will eventually have the lower pressure value. Finally the secondary takes over the circulation of the primary and one area of low pressure will then exist.

While secondary depressions may be frontal or non-frontal, the most common type are frontal which develop on the trailing cold fronts of primary depressions (Fig. 8.14(1)). Frontal secondary depressions have life cycles similar to that of any frontal depression and will be a member of a family of depressions, but all members of the family are not necessarily secondary depressions (compare Figs. 8.6 and 8.14(1)).

Non-frontal secondary depressions normally develop in the unstable air stream to the rear of an occluding depression. The form of the isobars associated with the feature is markedly circular (Fig. 8.14(2)). Cumuliform cloud and precipitation occur in its central area, and extend along the trough line on the side of the secondary furthest displaced from the primary depression. An area of slack pressure gradient with associated light variable winds exists between the two areas of low pressure. However, on the side of the secondary, furthest from the primary, there is an area of steep pressure gradient and high wind speeds.

ANTICYCLONES

The simplified general circulation implies the existence of high pressure areas in the subtropical and polar regions (Fig. 8.1). In the actual circulation these areas of high pressure take the form of anticyclones and are represented on surface charts by a set of closed isobars of oval or sometimes circular form (Figs. 8.3, 8.11). In contrast with frontal depressions, anticyclones move slowly eastwards, or may be stationary (Fig. 8.6).

In the sub-tropics, anticyclones persist for long periods and are termed *permanent*. Those in the North Atlantic are called the Bermuda and Azores anticyclones, reflecting their preferred location, and in the North Pacific there is a similar pattern of two anticyclones. However, in the southern hemisphere subtropical anticyclones move eastwards at greater speeds than those of the northern hemisphere.

The term *semi-permanent* is applied to the anticyclonic conditions which develop seasonally in temperate and polar zones. Examples are anticyclones which develop over North America and Siberia in winter, but it should be noted that these areas are also frequently affected by frontal depressions.

Anticyclones may be either *warm* or *cold*. Warm anticyclones (all subtropical and some higher latitude anticyclones) are typified by a mainly warm troposphere. A cold anticyclone is one in which the lower troposphere is occupied by cold air, thus it has a relatively shallow circulation (e.g. anticyclones which move rapidly into lower latitudes behind a family of depressions). Some anticyclones show both cold and warm attributes, with a shallow layer of cold air at and near the surface, and warm air in the troposphere above. These exist in temperate zones during the winter months, when slow moving or stationary warm anticyclones persist over a large land mass. The radiative cooling of the surface results in the development of a layer of cold air in the lower troposphere.

As the pressure gradient is slack, the central area of an anticyclone is either calm or has light and variable winds. Towards the outskirts of the system wind speed increases and wind direction becomes more marked, being outwards and *clockwise* in the *northern,* but *anticlockwise* in the *southern* hemisphere. The air flow is divergent at the surface, but convergent in the upper part of the troposphere with a subsequent downward movement of air in the troposphere, termed *subsidence* (Fig. 8.15(1)). In the early stages of the development of an anticyclone the rate of subsidence is high but decreases as the system develops. Subsidence is significant in modifying the environmental lapse rate as the air warms adiabatically. If the process continues over a period, an upper level temperature inversion develops, promoting a stable atmosphere at this level (Fig. 8.15(2)). The height at which the inversion is present varies within an anticyclone, and from one anticyclone to another, depending upon the degree of subsidence.

In the early stages of anticyclonic development, cloud in the upper and middle troposphere will clear as a result of subsidence, hence the clear skies associated with most anticyclones. Subsequently subsidence and adiabatic warming result in low relative humidity values, which are particularly noticeable at the level of the inversion.

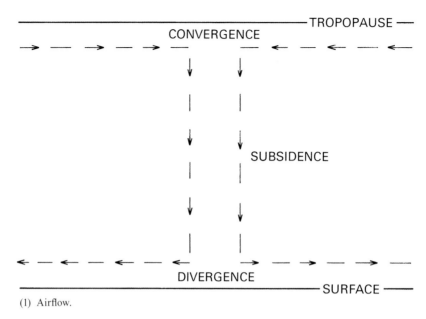

(1) Airflow.

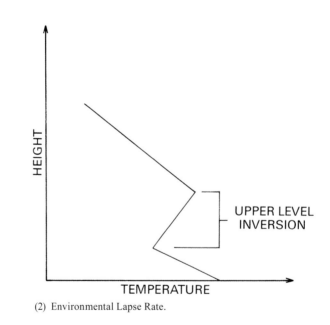

(2) Environmental Lapse Rate.

Fig. 8.15 Characteristics of an anticyclone.

Weather Conditions

In an anticyclone, the environmental lapse rate of the lower troposphere is variable, as it is affected by the diurnal variation of surface temperature. Such variations are most significant over land, and play an important part in determining anticyclonic weather conditions.

In summer months, a warm anticyclone over land results in high air temperatures during the day. Any cloud developing will be fair weather cumulus, its vertical extent being limited by the upper level inversion. Its life span will be short as the relative humidity of the surrounding air is low. Visibility on land may well be reduced as a result of haze, and in coastal areas sea breezes develop. Overnight clear skies and low wind speeds favour the formation of *dew,* a deposit of water droplets on the surface. Dew forms as a result of air, in contact with the surface, being cooled below its dew-point temperature. Radiation mist or fog may develop locally where there is a high relative humidity and downslope drainage of cold air. Soon after sunrise the mist or fog disperses.

During winter, the weather conditions initially are clear skies and low air temperatures during the day, while overnight air temperatures often decrease below freezing point. Condensation occurs if the air is cooled below its dew-point temperature. Water droplets forming on the surface may freeze and further cooling of the air results in sublimation and the development of a white crystalline deposit termed *hoar frost*. Alternatively, a high relative humidity and a long cooling period may cause widespread and rapid formation of radiation fog, which drifts offshore over coastal waters and estuaries. The fog may persist throughout the following day as a result of the small amount of solar radiation reaching the surface due to the time of year and the presence of the fog (Chapter 6). However, during the day the fog may lift over land to form a cover of low level stratiform cloud, which in the late afternoon appears to "lower" to the surface, the condition being termed *anticyclonic gloom*. If the air temperature is below 0°C, the fog droplets will be supercooled and will freeze on coming into contact with any object whose temperature is at or below 0°C. The white deposit of ice is called *rime* and always accumulates on the windward side of the object.

When smoke and other pollutants are trapped beneath the upper level inversion of an anticyclone because of the stable atmospheric conditions at this level, *smog* exists (Plate 26). This term is derived from smoke and fog, but is also used to describe conditions produced by pollutants without fog (e.g. in Los Angeles).

Anticyclonic weather conditions in summer and winter may persist for an extended period, and a change in conditions depends on the movement or general weakening of the anticyclone. The change is indicated by increasing wind speeds caused by steepening pressure gradients, normally associated with an advancing depression.

RIDGES OF HIGH PRESSURE

An extension of an anticyclone identified by the form of isobars on a surface pressure chart is termed a *ridge* (Fig. 8.3), the point of maximum curvature of the isobars denoting its axis. A ridge may often be a direct extension of a large anticyclone and produces typical anticyclonic weather while it persists. In contrast, a ridge located between two frontal depressions moves rapidly over an area, resulting in a short-lived break in the adverse weather of the depressions. If the axis of the ridge is approached from the east, decreasing wind speeds are experienced as the pressure gradient slackens. Cloud cover decreases, possibly giving clear skies, or cumuliform cloud without precipitation. To the west of the axis, wind speeds increase as the pressure gradient steepens. Cloud cover which increases and progresses from cirrus (Ci) to cirrostratus (Cs) indicates an advancing frontal depression.

COLS

If there are two anticyclones and two depressions arranged alternately, a feature exists between them termed a *col* (Fig. 8.3). Within the col the pressure gradient is slack, and the winds light and variable. Other weather conditions associated with a col will depend upon the effect of the underlying surface on the air mass above. Thunderstorms are probable if there is instability, but with stability radiation fog may develop, or advection fog where warm air passes over a colder surface. During winter, cols are relatively short-lived due to the rapid movement of depressions which surround them.

AIR MASSES

An air mass is a large body of air whose temperature and relative humidity values are more or less uniform in a given horizontal plane. Its horizontal dimensions are of the order of hundreds or even thousands of square kilometres. The most suitable conditions for the development of an air mass are a combination of a uniform part of the earth's surface and the presence of a large anticyclone for at least three days. The slack surface pressure gradient, and hence low wind speeds, of the anticyclone enables the air mass to acquire its properties. Thus the source regions of air masses are generally found in the subtropical, temperate and polar zones, where uniform surfaces and anticyclonic conditions exist simultaneously. Air masses are therefore classified as *Tropical, Polar* or *Arctic (Antarctic)*, to indicate their source regions and their relative temperatures. Arctic (Antarctic) air masses originate within the Arctic (Antarctic) circles, and are colder than the Polar air mass which originates to the south (north) of this region. Air masses are further sub-divided into *maritime* (high humidity), or *continental* (low humidity) as in the following table:

Air mass	Abbreviation	Source	Temperature	Relative humidity
Tropical maritime	Tm	Subtropical ocean areas e.g. Azores	High	Very High
Tropical continental	Tc	Subtropical deserts e.g. Sahara	Very high	Very low
Polar continental	Pc	Temperate continental area e.g. N. Europe	Varies with season	Low
Polar maritime	Pm	Ocean areas latitude $>50°$	Low	High
Arctic	A	Arctic ice cap	Very low	Low

Table 8.4 Characteristics of air masses.

Modification of Air Masses

Once an air mass leaves its source region, it will no longer be in equilibrium with the underlying surface which may be either warmer or colder than the air mass. The mass will be heated or cooled, and both its temperature and environmental lapse rate will change. If it is heated by the surface, the E.L.R. increases and it becomes unstable, whereas if it is cooled by the surface the E.L.R. decreases, and it becomes stable (Chapter 3). The relative humidity of the mass is also affected by changes in temperature, increasing when the mass is cooled, and decreasing when the mass is heated. However, when the mass is heated, evaporation from the surface occurs, counteracting the effect of heating on the relative humidity value. Therefore in establishing the final RH value of the air mass, both heating and evaporation effects must be taken into account.

Other processes in the atmosphere, for example convergence, which are not further considered, may also modify the characteristics of an air mass.

Weather Conditions

To illustrate the conditions associated with an air mass the weather experienced over the British Isles is taken as an example. This choice of example is not as restrictive as might first appear, since the conditions generated by an air mass are similar to those experienced in other parts of the world. Typical synoptic patterns associated with the five principal air masses affecting the British Isles are shown in Fig. 8.16. The nature of the surface over which each air mass passes before arrival over the British Isles is identifiable, thus changes in the characteristics of each air mass and the related weather conditions can be assessed.

Tropical maritime (Tm) affects the British Isles both in winter and summer. It moves from the Azores northwards over progressively cooler surfaces and, having been subjected to cooling, its relative humidity increases and it becomes stable. Advection fog may occur, particularly in spring and early summer, in the Western Approaches and English Channel. In the summer low level stratus may develop over the sea and also affect windward coasts. Inland, cloud amount is less, while on the east coasts clear skies are probable, reflecting the heating of the air mass as it passes over the warmer land surface.

During the winter months, advection fog is more probable over land, particularly if the surface has been subjected to a period of low temperatures. Dependent upon the wind speed, low level stratus with drizzle is possible over land or sea. Orographic cloud with rain may develop over high ground. Air temperatures are above the seasonal average.

Tropical continental (Tc) is uncommon, and is only experienced over the British Isles in the summer months. From its source region in the Sahara desert it passes over the Mediterranean and Europe. Although cooled slightly, it produces high air temperatures well above the seasonal average. During the day unstable conditions and convection currents develop. However, in spite of its sea passage the relative humidity of the air mass is low, thus there is a general lack of cloud, and the air temperature is enhanced. Haze may develop.

Polar continental (Pc) originates in Northern and Western Europe and Russia. During the winter, the North Sea is warmer than the air mass and instability develops as the air is heated, its relative humidity simultaneously increasing. Convection over the sea may result in cumulus and cumulonimbus clouds, but, if the sea passage is short, only stratocumulus may develop. The eastern seaboard is affected by cloud with showers of snow, hail or sleet. Western areas experience generally clear sky conditions which favour the development of hoar frost, or possibly radiation fog overnight if wind speeds are low enough. Below seasonal average temperatures are experienced in spite of the heating of the air mass.

In summer the Polar continental air mass is cooled by the North Sea and becomes more stable. Advection fog or low level stratus forming over the sea affects the east coast, where the air temperatures are generally below the seasonal average. In contrast, areas a short distance inland and to the west experience clear skies with higher air temperatures, and haze may develop during the afternoon.

Polar maritime (Pm) is an air mass frequently affecting the British Isles. Approaching from the west and north-west, its maritime properties of high relative humidity and unstable conditions develop during its long transit over the North Atlantic. Convection over the sea results in cumulus and cumulonimbus clouds, which in winter on the western side of the country are accompanied by showers of rain, sleet or snow or sometimes hail. Eastern areas experience little or no cloud cover, thus favouring the development of hoar frost or radiation fog during the night. Air temperatures are generally the seasonal average. In summer convection occurs over sea and land, and cloud cover is widespread with associated showers of rain, hail, and possibly thunderstorms. The air mass is cool with air temperatures below the seasonal average.

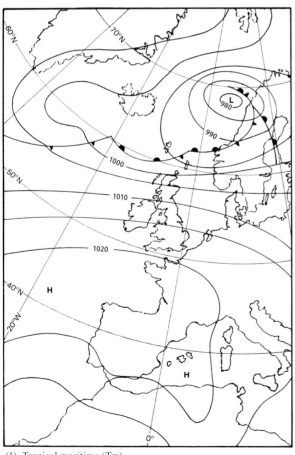

(1) Tropical maritime (Tm).

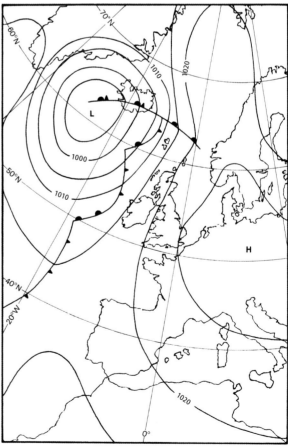

(2) Tropical continental (Tc).

Fig. 8.16 Air masses – typical surface synoptic chart.

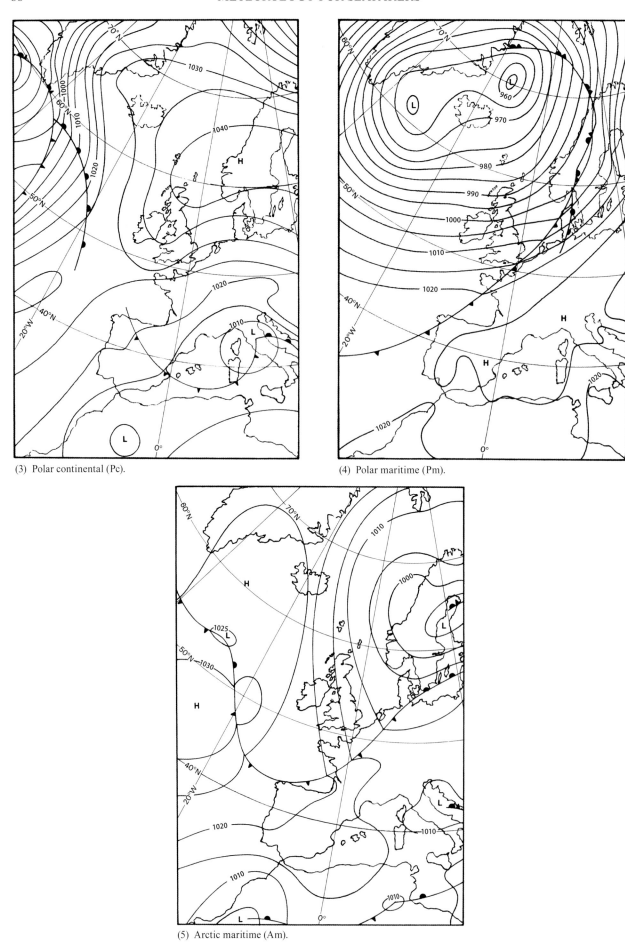

(3) Polar continental (Pc).

(4) Polar maritime (Pm).

(5) Arctic maritime (Am).

Fig. 8.16 Air masses – typical surface synoptic chart.

Arctic (A) is an air mass originating in the Arctic Basin, and is normally experienced in winter and spring and occasionally in the other seasons. On its passage to the British Isles, maritime properties develop and it becomes *Arctic Maritime* (Am). Its relative humidity increases, and a steep lapse rate develops as it is heated by the warmer sea. Showers of snow and hail are usually experienced in Scotland and on the windward coasts of the North Sea. Skies are clear inland, and air temperatures are below the seasonal average.

CHAPTER 9

TROPICAL AND SUBTROPICAL CIRCULATION

INTRODUCTION

The tropical zone spans the area from the Equator to 23° 27' N and S, and the subtropical zones from 23° 27' N and S to approximately 40° N and S Surface synoptic charts of these zones show significant differences compared with those of the temperate zones, the pressure distribution appearing generally less complex with slacker pressure gradients (Fig 9.1). In addition to subtropical anticyclones (Chapter 8), tropical cyclones, the Intertropical Convergence Zone (ITCZ), Trade Winds and monsoons are features of the atmospheric circulation in these zones.

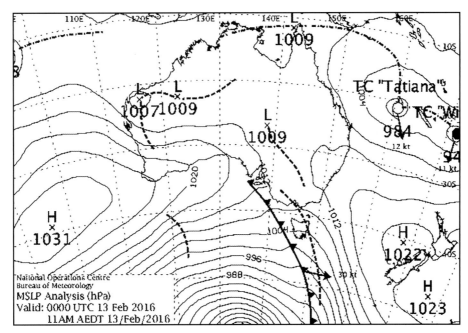

Fig. 9.1 Tropical and sub-tropical circulation–typical surface synoptic chart.

TROPICAL CYCLONES

Terminology

A *tropical cyclone* is an area of low pressure which develops over tropical or sub-tropical waters and is delineated on a surface chart by a number of closed and almost perfectly circular isobars, with associated steep pressure gradients. It is also non-frontal.

In development, a tropical cyclone can usually be traced through the stages of a tropical depression and storm, each system being differentiated by its maximum sustained wind speeds as follows:–

(a) *Tropical depression (TD)* – Not exceeding Beaufort Force 7.

(b) *Moderate tropical storm (TS)* – Beaufort Force 8 and 9.

(c) *Severe tropical storm (STS)* – Beaufort Force 10 and 11.

(d) *Tropical cyclone* – Of at least Beaufort Force 12 (32.7 ms⁻¹)

On a surface chart the centre of a tropical cyclone may be denoted by the symbol ◖ and a tropical storm by ◖ for systems in the southern hemisphere (Fig. 9.1) and respectively ◗ and ◗ for the features in the northern hemisphere.

Tropical cyclones have been known historically by a number of local names, some of which are currently used in weather bulletins issued by meteorological services e.g. *typhoon* in the Western North Pacific (Fig. 9.2) and *hurricane* in the North Atlantic and Eastern North Pacific. For the Arabian Sea, Bay of Bengal, South Indian and Western South Pacific Oceans the term is *cyclone*, however, a check is recommended of the definitions of the terms used by each meteorological service. Each system is allocated a name when it becomes a tropical storm (Figs. 9.1 and 9.2). The maritime term for a tropical cyclone is a *tropical revolving storm*.

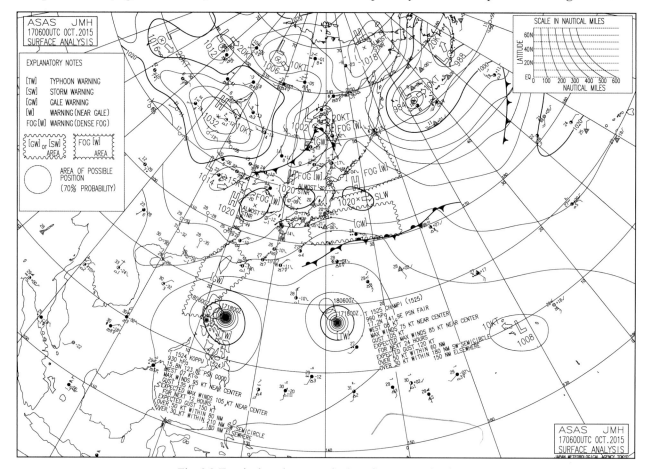

Fig. 9.2 Tropical cyclone – typical surface synoptic chart.

Distribution

Tropical cyclones develop in eight clearly defined areas during specific seasons with the average number per annum varying from one area to another (Table 9.1 and Fig 9.3). However, the average numbers shown in Table 9.1 should be taken as a guide, and may in any year be significantly different.

Area	Northern Hemisphere												Av no/yr
	Jan.	Feb.	Mar.	Apr.	May	Jun.	Jul.	Aug.	Sep.	Oct.	Nov.	Dec.	
North Atlantic Ocean						HURRICANES							8
Eastern North Pacific Ocean					HURRICANES								7-8
Western North Pacific Ocean					TYPHOONS								22
Bay of Bengal					CYCLONES								5-6
Arabian Sea				CYCLONES									1-2
Area	Southern Hemisphere												
Western South Indian Ocean	CYCLONES												6
Eastern South Indian Ocean	CYCLONES												1
Western South Pacific Ocean	CYCLONES												2-3

Table 9.1 Tropical cyclone seasons.

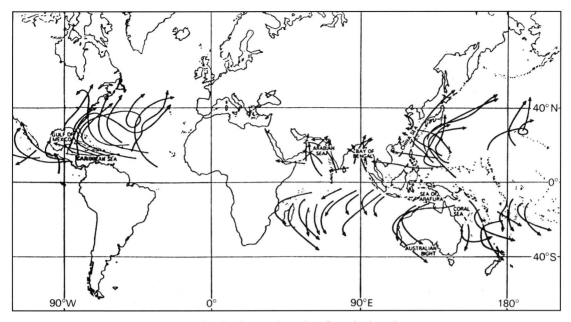

Fig. 9.3 Distribution and tracks of tropical cyclones.

Structure

A tropical cyclone has an average diameter of 740 km and an average central pressure value of 960 hPa at mean sea level (870 hPa has been recorded), with very steep pressure gradients. However, during its life span both diameter and pressure values vary. The system has a distinctive eye, or vortex, of some 55 km in diameter occupying its central area. In this area calm conditions or light airs exist at the surface but, immediately outside, winds are Beaufort Force 12 or greater. The eye is generally free from cloud since within it the air subsides and warms adiabatically. The eye is surrounded by the eye wall, which is a ring of cumulonimbus clouds formed by the air ascending (Fig. 9.4). This sharp contrast in cloud conditions is often illustrated by satellite images (Fig. 9.5). On the periphery, there are spiral banks of cloud, whose vertical development and horizontal area are not as great as that immediately outside the eye (Fig. 9.4). In satellite images the anvils of the Cb cloud in the central area may often overshadow the peripheral clouds.

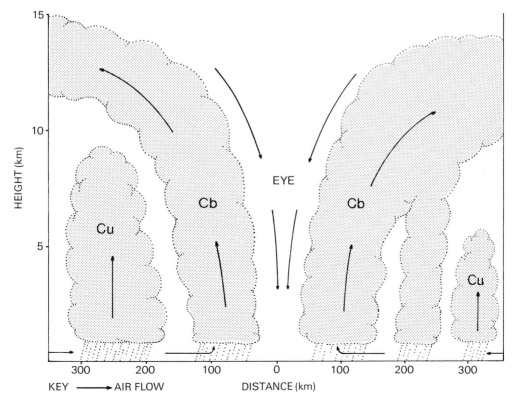

Fig. 9.4 Tropical cyclone elevation (For clarity the diameters of individual cloud cells are exaggerated).

Fig. 9.5(1) Tropical cyclone – geostationary satellite visible image.
Severe Tropical Storm Koppu and Typhoon Champi – Himawari 8 – 0300UTC 19th October 2015.

Fig. 9.5(2) Tropical cyclone – geostationary satellite infra-red image.
Severe Tropical Storm Koppu and Typhoon Champi – Himawari 8 – 0300UTC 19th October 2015.

Formation

The more frequent formation of tropical cyclones during summer and autumn in particular areas suggests that certain conditions are necessary. Research has shown that a sea surface temperature of at least 27°C is required. At this temperature saturated air has a large water vapour content (Fig. 4.2 – T), and as the air ascends adiabatic cooling causes condensation with the release of latent heat of vaporization, which is the source of energy for the development and continued maintenance of a tropical cyclone. The significance of the sea surface temperature is also illustrated by the point of origin of a cyclone tending to move into higher latitudes and subsequently into lower latitudes as the season progresses. The tracks of tropical cyclones (Fig. 9.3) indicate that formation occurs at least 5°N or S of the Equator, as the value of the Coriolis force is critical in the development of the surface winds in the low pressure system.

In an atmosphere where the vertical wind shear has a small value and preferably one where the direct influence of a jet stream is absent, the cumuliform cloud in a tropical depression can develop great vertical extent. With the large amounts of latent heat of vaporization released on condensation of water vapour within the developing clouds, the system develops a warm core. These factors contribute to the decrease of the central pressure value, with the attendant steepening pressure gradient and increase in surface wind speeds, thus further energy enters the system. The eye of the future tropical cyclone first appears in the upper part of the troposphere and then descends. As the air in the eye descends (Fig. 9.4), it warms adiabatically, enhancing the development of the warm core. If favourable atmospheric and sea surface conditions persist, the system continues to evolve, associated wind speeds increase, and it becomes a tropical cyclone. Of the total energy entering the system, a small amount is converted into kinetic energy, but a far larger amount is exported through its upper level circulation associated with the Cb anvils.

The period of development from a tropical depression through to a cyclone is by no means consistent from one system to another, and may range from explosive development of less than 24 hours duration to one of several days (Fig. 9.6). In weather bulletins development may be identified by the system being *upgraded* for example from a moderate tropical storm to a severe tropical storm, when the maximum sustained surface wind speed has increased.

Decay

During its evolution a tropical cyclone may move onshore, and, as the energy derived from water vapour is reduced, the tropical cyclone may decay rapidly. If it subsequently moves offshore it may well be revitalised. Whether a tropical cyclone moves out to sea or remains offshore throughout its existence, it may eventually move into higher latitudes. There, with the gradual reduction in sea surface temperature and the probability of increasing vertical wind shear, the central pressure value of the system increases and the surface wind speeds decrease. With decreasing surface wind speeds weather bulletins may state that the system has been *downgraded*. In higher latitudes the warm air of the system may become part of the warm sector of a frontal depression and is now identified as being extratropical (Fig. 9.6).

Movement

The initial track of a tropical cyclone in lower latitudes is generally from east to west as it is steered by the easterly tropospheric air flow in which it is embedded. As a pole-ward component is also present, its track may typically be close to WNW in the northern hemisphere and WSW in the southern hemisphere, along the equatorward margin of the subtropical anticyclone. During this stage the system moves at about 10 knots. When a tropical cyclone reaches the western edge of the subtropical anticyclone, usually between 20° and 30°N and S it may recurve. When recurving, the initial track of the cyclone is north (N Hemisphere) or south (S Hemisphere), its speed of advance decreases, and it may become stationary for a period. Subsequently it tracks north-easterly (N Hemisphere) and speed of advance increases to 20 knots or greater. The system has now left the tropical circulation and is being steered by the general westerly tropospheric airflow of the mid-latitudes (Fig. 9.3)

On occasions a tropical cyclone may not recurve, but continue on a path with a westerly component. In most cases this is directly related to the position and movement of the subtropical anticyclone which blocks the path of recurvature. At a later stage the system may recurve unless it moves onshore and decays rapidly.

The variable paths of tropical cyclones are illustrated by the four examples in Fig. 9.6. All the systems developed in 2015, commencing with Typhoon Mujigae on September 30th, whose path had a distinct westerly component and simultaneously moved into higher latitudes, but did not recurve (Fig. 9.6(1)). In contrast Severe Tropical Storm Choi-wan (Fig. 9.6(2)) was clearly commencing recurvature on October 6th, but was unable to complete the process as a tropical system, becoming extratropical on October 7th.

Key to Tropical Cyclone Tracks

ExT – Extratropical   STS – Severe tropical storm   TD – Tropical depression   TS – Moderate tropical storm   TY – Typhoon

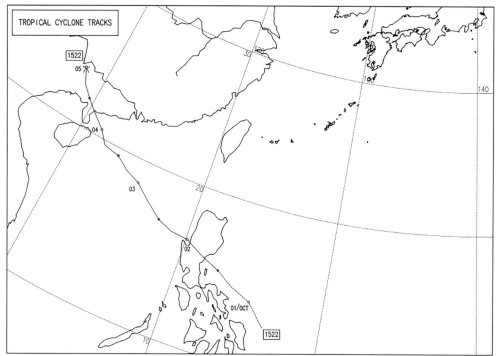

Typhoon Mujigae		
Date	Time (UTC)	Status
30.09.15	1800	TD
01.10.15	1200	TS
02.10.15	1200	STS
03.10.15	1200	TY
04.10.15	1800	STS
05.10.15	0000	TD

Fig. 9.6(1) Tropical cyclone track – Typhoon Mujigae

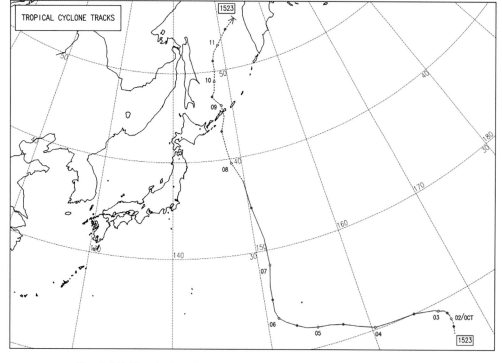

STS Choi-wan		
Date	Time (UTC)	Status
01.10.15	0600	TD
02.10.15	0600	TS
05.10.15	0000	STS
07.10.15	1800	ExT

Fig. 9.6(2) Tropical cyclone track – Severe Tropical Storm Choi-wan

For the system which developed into Typhoon Koppu (Fig. 9.6(3)), its initial track was on a westerly heading covering approximately 30° of longitude before recurvature commenced on October 18th. However, the system was unable to progress any significant distance into higher latitudes due to its subsequent position relative to the subtropical anticyclone (Fig. 9.7). In comparison Typhoon Champi (Fig. 9.6(4)) moved on a more typical path for a tropical cyclone. Its position relative to the subtropical anticyclone and the southern boundary of the mid-latitude circulation allowed the system to commence recurvature during October 16th. In the subsequent days Champi showed a marked decrease in speed of advance, but on October 21st speed of advance increased as it moved to an easterly heading.

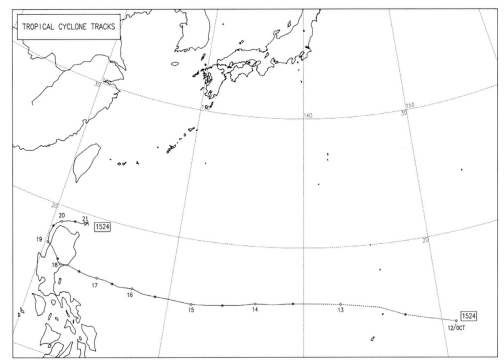

Typhoon Koppu		
Date	Time (UTC)	Status
12.10.15	0000	TD
13.10.15	1200	TS
15.10.15	0600	STS
15.10.15	1800	TY
18.10.15	1800	STS
19.10.15	0600	TS
21.10.15	0600	TD

Fig. 9.6(3) Tropical cyclone track – Typhoon Koppu

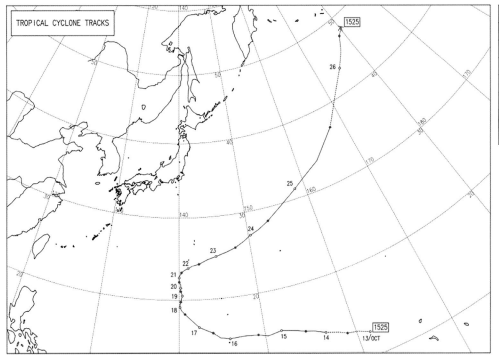

Typhoon Champi		
Date	Time (UTC)	Status
13.10.15	0000	TD
14.10.15	0000	TS
15.10.15	1800	STS
17.10.15	0000	TY
24.10.15	0000	STS
25.10.15	1200	ExT

Fig. 9.6(4) Tropical cyclone track – Typhoon Champi

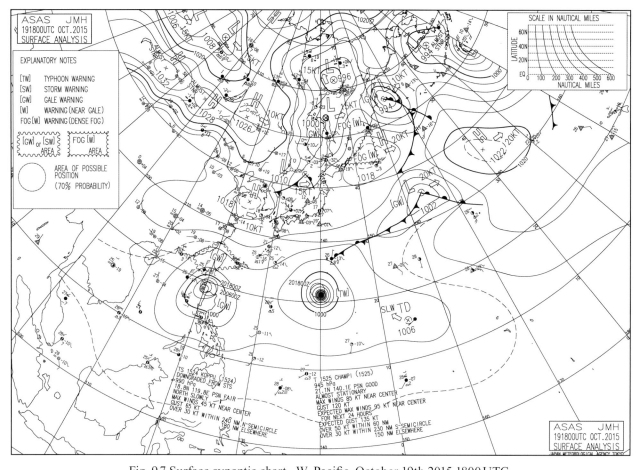

Fig. 9.7 Surface synoptic chart – W. Pacific, October 19th 2015 1800 UTC.

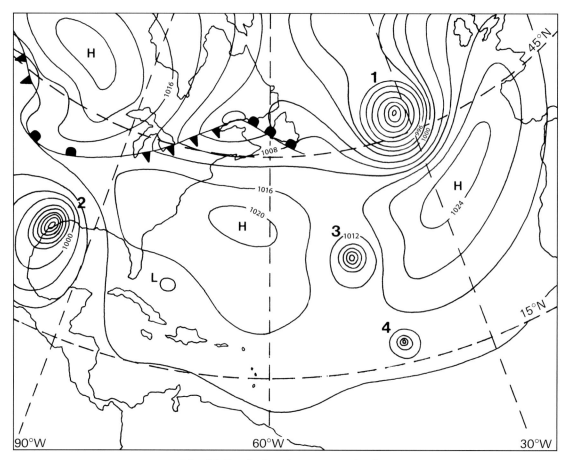

Fig. 9.8 Tropical cyclones – N. Atlantic.

Fig. 9.8 shows examples of tropical cyclones for the North Atlantic Ocean, where the movement of 2 was influenced by a subtropical anticyclone and it did not recurve, whereas the preceding system 1 had recurved along the eastern seaboard of North America. In mid–Atlantic, System 3 is recurving and 4 may follow the same path

As no two tropical cyclones have exactly the same track, mariners are advised to monitor carefully the movement of each system. To assist mariners, meteorological services frequently issue and update bulletins and charts which may include an element of uncertainty as to the future movement of a system. Figs 9.2 and 9.8 illustrate one style of presentation, where for each system the probable future positions of the centre for a specified time and date is shown by a circle.

An alternative is the delineation of a danger area also known as an area of avoidance (Fig 9.9). Three factors are considered in the construction of the danger area; the forecast 72 hour movement of the centre of the cyclone shown by the symbols ◗ , possible error in the 24, 48 and 72 hour forecast movement, which increases respectively in equal increments from 100 to 300 n.miles, and the outer limit of the area likely to be affected by winds of 34 knots for each forecast time.

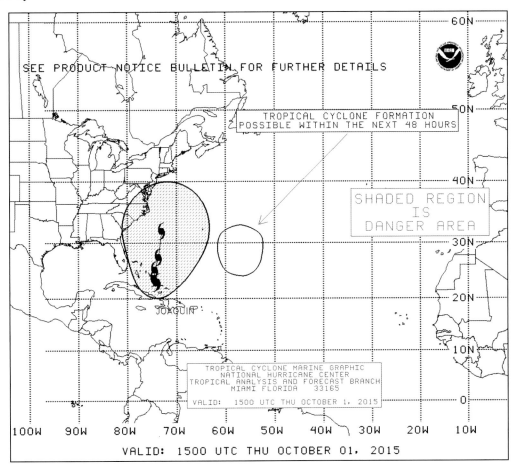

Fig. 9.9 Tropical cyclone marine graphic – October 1st 2015 1500 UTC.

Weather and Sea State

Weather and sea state conditions associated with tropical cyclones are extreme. The loss of life and financial costs arising from the damage caused can be high. Fig. 9.10 shows the pressure trends and probable wind conditions likely to be encountered in an average tropical cyclone. The sea state is directly related to these conditions, while within the eye the seas are confused due to the high winds which recently prevailed. Extreme storm waves generated by the wind may be amplified by astronomical tides and thus become storm surges, which cause extensive damage in coastal areas (Appendix 1). Precipitation will be rain showers, light on the periphery of a cyclone and increasing in frequency and intensity, to reach a maximum on the outskirts of the eye itself. Within the eye low level stratiform cloud may occur over the sea and clear skies over land.

Warning Signs

Given the adverse conditions associated, the mariner should avoid entering a tropical cyclone. In order to determine the best course to follow, weather bulletins and charts, issued by meteorological services, need to be carefully analysed and collated (Figs 9.7 and 9.9). In addition to this forward assessment, it is most important for the seafarer to observe at frequent intervals the weather and sea state, as these may provide vital warning signs of the presence of a tropical cyclone, should the vessel approach and enter the system. The warning signs are:–

Swell: Depending upon the extent of the open ocean, an early indication of a tropical cyclone is the swell generated by the winds in the systems. These swells, whose general characteristics are long wavelengths and low wave height have been observed at some 600 n.miles from the centre. The direction from which the swell comes is an indication of the bearing of the centre of the cyclone when the swell was generated.

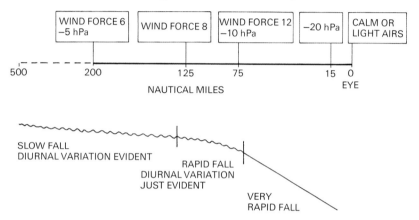

Fig. 9.10 Tropical cyclone – wind and pressure distribution.

Skies: An unusually clear day may precede the appearance of cirrus cloud with brightly coloured sunsets and sunrises. The cloud base subsequently lowers, with the onset of showers, in the outer region of the system.

Atmospheric pressure: A decrease in atmospheric pressure can indicate the presence of a tropical cyclone. The typical diurnal variation of pressure in low latitudes is evident close to and within the area of the system, but this is superimposed on an overall decrease in pressure (Fig 9.10). The following procedure is recommended to determine whether or not a tropical cyclone is present:

(a) Read the barometer and correct the reading to standard datum and, if applicable, for index error (Chapter 2).

(b) Correct (a) for diurnal variation of pressure from table in pilot books or climatological atlases, taking into account latitude and local mean time.

(c) Compare corrected reading (b) with mean pressure for the time of year. If the corrected reading is 3 hPa below the mean pressure there is a possibility of a tropical cyclone. If it is 5 hPa below, then a tropical cyclone must be assumed to be present with the vessel on the outskirts of the system (Figs. 9.10 and 9.11).

Wind: An increase in wind strength and possibly a change in direction.

Single Observer's Action

If a vessel enters a tropical cyclone, frequent and careful monitoring of the warning signs will give the seafarer time to decide on action to be taken. Termed the *Single Observer's Action,* it is set in the context of the maritime diagrammatic representation of a tropical cyclone Fig. 9.11. Each of the three circles shown is a decrease in pressure below mean pressure value (cross reference with Fig. 9.10). Surface wind direction is shown and the angle of indraught decreases progressively from the periphery of the system to zero at the outskirts of the eye. The terms in Fig 9.11 are defined in Table 9.2 and, although originally devised in the days of sail, are still relevant to the handling of power-driven vessels.

Alternative terms which may be used for the dangerous and navigable semicircles are right and left hand semicircles. The term which applies to a semicircle assumes the observer is looking in the direction towards which the cyclone is moving. In the northern hemisphere the dangerous semicircle would be the right hand semicircle.

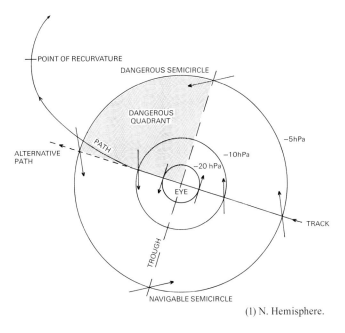

(1) N. Hemisphere.

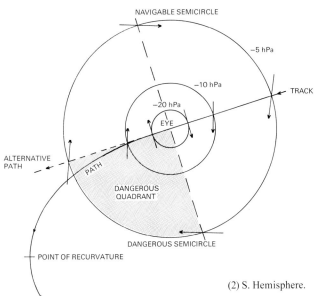

(2) S. Hemisphere.

Fig. 9.11 Tropical cyclone – surface plan view.

Eye or Vortex	Central area of calm, surrounded by winds of Force 12 or greater.
Track	The positions over which a tropical cyclone's centre has already passed.
Path	Anticipated movement of the tropical cyclone.
Trough	The line of lowest barometric pressure which would be noted by stationary observer.
Bar (Eye wall)	The mass of clouds surrounding the eye.
Dangerous semicircle	The semicircle where a vessel should not run before the wind, as it may enter the path and the eye.
Dangerous quadrant	The advance quadrant of the dangerous semicircle wherein the tropical cyclone may recurve over a vessel.
Navigable semicircle	The semicircle where a vessel may run before the wind, and thus move away from the eye.
Point of recurvature (Vertex or Cod)	The point on the track of a tropical cyclone where the curvature is greatest.

Table 9.2 Tropical cyclone – Maritime terminology.

The Single Observer's Action incorporates the following:

(a) Establish the position of the vessel in the tropical cyclone.

As the changes of both wind direction and speed over a period of time are the most significant indicators of the vessel's position, it is recommended that the vessel heaves to for two to three hours to observe the wind. The changes which would indicate the vessel's position are as follows:

Wind	Vessel's position	
	Northern hemisphere	Southern hemisphere
Veering	Dangerous semicircle	Navigable semicircle
Backing	Navigable semicircle	Dangerous semicircle
No change in direction but increase in speed	Vessel is in the path	

Atmospheric pressure decreasing (increasing) will indicate whether the vessel is in advance (behind) the trough of the system

(b) Determine the track of the tropical cyclone.

While hove to, the bearing of the centre of the cyclone can be determined by applying Buys Ballot's Law. If the observer faces the wind, then the bearing will be to the right (left) in the northern (southern) hemisphere as follows:

Pressure below mean	5 hPa	10 hPa	20 hPa
Bearing	12 points	10 points	8 points

NB: 1 point = 11.25°

To determine the position of the centre relative to the vessel, the wind speed and decrease in pressure below the mean, also need to be considered, these being based on the "average" tropical cyclone (Figs 9.10 and 9.11). The approximate track of the cyclone can be established from two or more bearings of the centre, with an interval of two to three hours between each observation.

(c) Evasive Action

Once the vessel's position in the tropical cyclone is established, evasive action as indicated in Figs. 9.12 and 9.13 can be taken using best speed. The recommended heading of the vessel with respect to the wind should increase the distance between the vessel and the centre of the cyclone, and with the alteration of course the vessel will move in a direction opposite to that of the cyclone's path. When the vessel is behind the trough the atmospheric pressure will increase. While evasive action is being taken, the seafarer must continue to monitor the conditions as these will indicate when the vessel is clear of the cyclone.

In latitudes where recurvature is possible, monitoring is particularly important, since if a vessel in the dangerous semicircle does not alter course in response to the change in wind direction, re-entering the cyclone and encountering the eye becomes a real possibility.

Where insufficient sea room prevents the evasive action recommended, a vessel in the dangerous semicircle should if possible heave to with the wind on the starboard (port) bow in the northern (southern) hemisphere, thus heading away from the centre. In the navigable semicircle in either hemisphere, the vessel should heave to in the most comfortable position relative to the wind and sea conditions.

The regulations in the International Convention for the **Safety of Life at Sea 1974 (SOLAS 1974)** recommend that vessels encountering a tropical cyclone should inform the nearest coastal radio station and all other vessels in the vicinity of its existence, and should update the information at least every three hours while it is affected by the system.

Appendix I is an extract of an account of a particular tropical cyclone reproduced by permission of Mr J. W. Nickerson and the *Mariners Weather Log* Vol 27 No. 1. The additional comments were published in *Seaways*, May 1984.

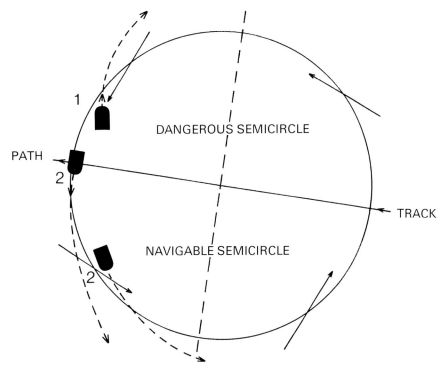

	Semicircle	Evasive action
1	Dangerous	Place wind 1–4 points on starboard bow altering course to starboard as wind veers.
2	Navigable or path of storm	Place wind on starboard quarter altering course to port as wind backs.

Fig. 9.12 Evasive action by vessel in N. Hemisphere.

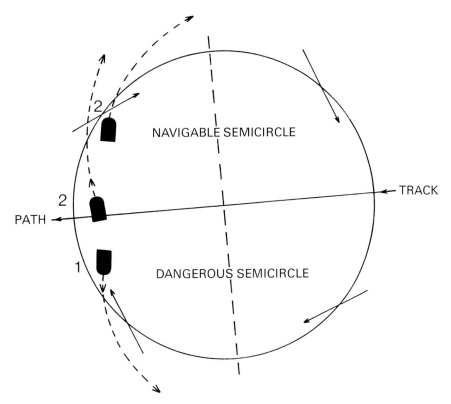

	Semicircle	Evasive action
1	Dangerous	Place wind 1–4 points on port bow altering course to port as wind backs.
2	Navigable or path of storm	Place wind on port quarter altering course to starboard as wind veers.

Fig. 9.13 Evasive action by vessel in S. Hemisphere.

INTERTROPICAL CONVERGENCE ZONE

The *Intertropical Convergence Zone* (ITCZ) is the area in low latitudes in which the *Trade Winds* of the two hemispheres converge. In this zone the horizontal convergence and subsequent ascent of air in the troposphere results in cloud development. Analysis of satellite images show that while cloud development is a feature of the zone, it may not be continuous in the horizontal plane at any given time, nor is it a persistent feature at any particular point over a time period (Fig. 9.14). The cloud type is cumuliform, ranging from cumulus to cumulonimbus, with diameters between 1 and 10 km. A collection of these clouds, termed a *convective cell,* may have a diameter between 10 and 100 km. A number of such cells may form a *cloud cluster* which may have a diameter between 100 and 1000 km. Weather conditions may be light winds, squalls and showers (Cu cloud), or thunderstorms and heavy precipitation (Cb cloud). In contrast clear skies indicate that there is no convergence of air with its resultant ascent.

On a surface chart the ITCZ can be shown by the pair of lines linked with cross hatching as illustrated on Fig. 9.15.

Fig. 9.14 Intertropical Convergence Zone – geostationary satellite infra-red image.
GOES 13 – 0000 UTC 9th June 2016.

The intermittent nature of convergence in the ITCZ, both in time and space, has resulted in the use of an alternative term *Intertropical Confluence*. Confluence implies the nearer approach of adjacent streamlines in the direction of the airflow; a *streamline* being a curve plotted parallel to the instantaneous direction of the wind vector (Fig. 9.16). The technique of streamline analysis is of great value in tropical meteorology since areas of surface convergence may be defined. The convergent flow may be that associated with a tropical depression which, under suitable conditions, may later develop into a tropical cyclone. Fig. 9.16 illustrates two fully developed tropical cyclones and one tropical storm.

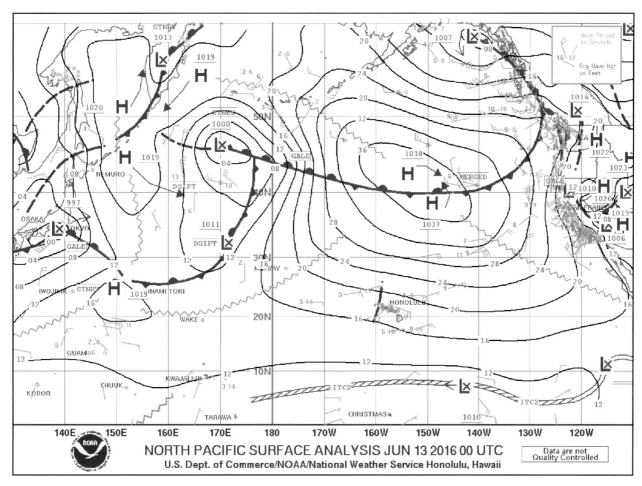

Fig 9.15 Intertropical Convergence Zone – Surface synoptic chart June 13th 2016 0000 UTC

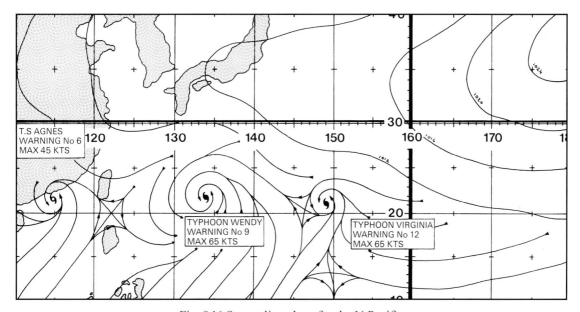

Fig. 9.16 Streamline chart for the N Pacific.

The position of the ITCZ varies throughout the year (Fig. 9.17), progressing north between January and July, and south in the latter half of the year. The range of latitude covered varies from one longitude to another, being greatest in the Indian Ocean, and least on the eastern side of the North Atlantic and North Pacific Oceans, where it does not move south of the equator. In the South Atlantic its progress south of the Equator is minimal, a factor which may partially account for the lack of tropical cyclones in this ocean (Fig 9.3). In certain ocean areas the ITCZ may be a single band (Fig. 9.17), or it may be a more complex double band feature.

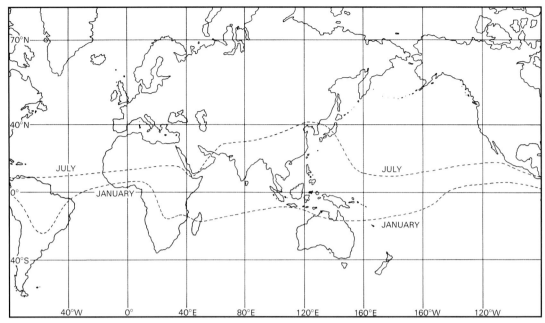

Fig. 9.17 Intertropical Convergence Zone – mean position in January and July.

Two terms which may be linked with the ITCZ are the *Equatorial Trough* and the *Doldrums*. The Equatorial Trough is defined as shallow trough of low pressure in which there is convergence of air moving equatorwards, the *Trade Winds,* from the subtropical anticyclones. As with the ITCZ, the trough migrates during the year over a range of latitude. This term may appear to be synonymous with the ITCZ; however, analysis of convergence in this area of the atmosphere shows that it occurs more frequently outside and on the side of the trough nearest the Equator.

Doldrums is the term for the zone in the equatorial ocean regions where light and variable winds occur. The area of the Doldrums varies seasonally both in latitude and longitude (Fig. 9.18), and is particularly well defined in the Eastern Atlantic and Pacific regions. Although the Doldrums are generally known for their light and variable winds, squalls, heavy rain and thunderstorms may well be experienced. A comparison of Figs. 9.17 and 9.18 will indicate where and at what time of the year the doldrums and the ITCZ are coincident.

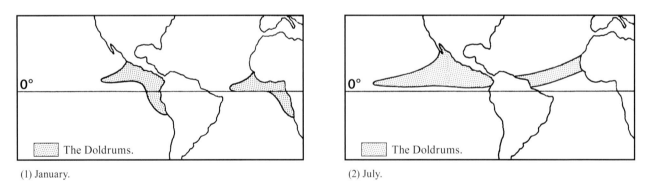

(1) January.   (2) July.

Fig. 9.18 The Doldrums – mean position.

TRADE WINDS

Trade winds are defined as the winds which exist between the subtropical anticyclone and the ITCZ. The North-East Trades are in the northern hemisphere and the South-East Trades in the southern hemisphere. The term derives from their reputation for constant direction and speed, which made them reliable for sailing vessels. Analysis of the pressure distribution on surface charts e.g. Figs 9.1, 9.15 and 9.19 will indicate where Trade Winds will exist over the oceans. Globally, the general extent of Trade Winds, as shown by the vector mean winds, the mean position of the ITCZ, and mean pressure value for the subtropical areas are illustrated for January in Fig. 9.20(1) and (2) and for July in Fig. 9.21(1) and (2).

It should be noted that in the North Indian Ocean in July the typical North-East Trades evident in other areas are absent, and the particular feature of the area is the *Monsoon* (see below).

The wind direction within each trade wind zone varies, and steadiness of direction and speed is a marked feature only in the core of the zone in each area. The average strength is Force 4, which can vary in the short term. There is a general tendency for an increase in strength during the winter season in each hemisphere. The cloud is trade wind cumulus which occurs as cloud streets, with an average cover of four-eighths (Plate 27). An important cloud characteristic is its greater vertical development and associated showers nearer the equator and western side of the oceans, reflecting the height at which the trade wind inversion exists in the troposphere. On occasions, the Trade Winds may be absent, their place being taken by other systems e.g. tropical depressions or tropical cyclones, Figs. 9.1 and 9.16.

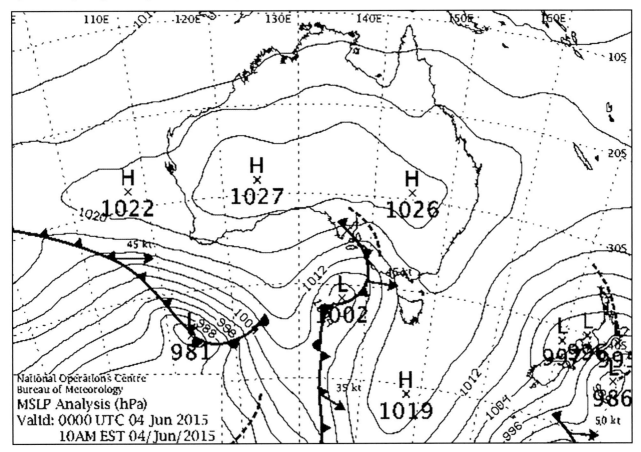

Fig. 9.19 Trade Winds and subtropical anticyclones – Surface synoptic chart June 4th 0000 UTC 2015

MONSOONS

Monsoon, derived from the Arabic "mausim" meaning season, was the name originally given to the winds in the Arabian Sea which reversed in direction from one season to another. It is now used to define that part of the atmospheric circulation covering a large geographical area over which a clearly predominant wind direction prevails in one season, with a reversal or near reversal of wind direction in the following season. The phenomenon occurs over the North Indian Ocean and most of the Southern Asian continent, where the North-East Monsoon prevails in winter months (Fig 9.22), and the South-West Monsoon, in the summer months (Fig. 9.23). Regional features such as the Tibetan Plateau and the Himalayas play a significant part in their generation.

In the North Indian Ocean the monsoon periods are as follows:

North-East Monsoon – November to March

Intermonsoon period – April and May

South-West Monsoon – June to September

Intermonsoon period – October

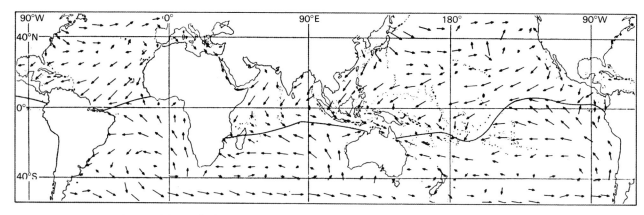

Fig. 9.20(1) Tropical and subtropical zones – January, ITCZ and the Trade Winds.

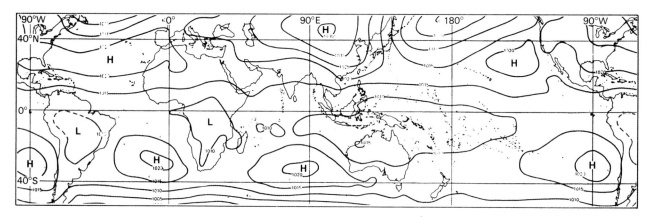

Fig. 9.20(2) Tropical and subtropical zones – January, mean pressure distribution.

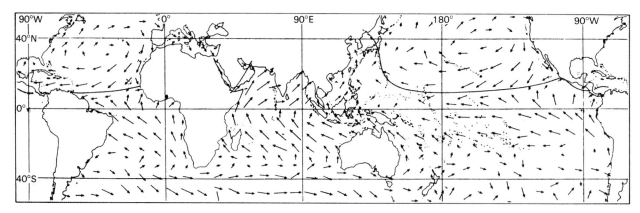

Fig. 9.21(1) Tropical and subtropical zones – July, ITCZ and the Trade Winds.

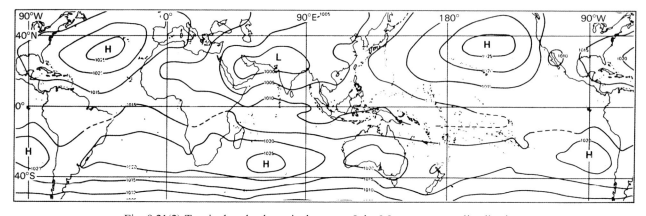

Fig. 9.21(2) Tropical and subtropical zones – July, Mean pressure distribution.

North-East Monsoon

The mean pressure chart for January (Fig. 9.20(2)) shows an extensive area of high pressure over the Asian continent, with decreasing pressure south towards the Equator. The vector mean wind chart (Fig. 9.20(1)) for the same month, shows north-easterly winds across the Arabian Sea and the Bay of Bengal. The typical surface synoptic chart for a day in January shows in greater detail the actual surface pressure distribution (Fig. 9.22 and Fig. 9.24), and confirms the north-easterly winds. At the height of the season the winds are Force 3 to 4 in the area, except near the Equator where Force 2 to 3 is normal. Generally, cloud cover is minimal with little or no rain. In December and January there is more cloud and rain in the southern area of the Bay of Bengal. The monsoon may be interrupted in the Arabian Sea north of 20°N, with wind direction changing and speed increasing when a frontal depression passes eastwards across the Asian continent.

South-West Monsoon

The mean pressure chart for July (Fig. 9.21(2)) shows an extensive area of low pressure centred over North-West India, with increasing pressure values towards the Equator. The vector mean wind chart (Fig. 9.21(1)) for the same month shows south-west winds over the North Indian Ocean, which corresponds to this pressure distribution. The typical surface synoptic chart for a day in July (Fig. 9.23) confirms this distribution. The area of low pressure over North-West India can be termed a *thermal low*, and in addition, other synoptic features can be present, which result in rain over the Indian subcontinent and the surrounding sea areas.

At the height of the monsoon, winds over the Arabian Sea achieve Force 6, with Force 7 or greater being possible on more than ten days per month. To the east of Socotra such conditions occur for 50% or more of the time in July. In the Bay of Bengal winds average Force 4 to 5, with Force 7 or greater occurring between 5 to 10 days a month in July and August. In the area between 5°N and the Equator and east of 60°E winds are generally Force 3, their direction being between south and west.

Off the west coast of India *offshore troughs* develop and result in periods of heavy rain, and the Western Ghats contribute to orographic precipitation. Further north in the Arabian Sea, *mid-tropospheric cyclones* can result in extensive cloud and precipitation for several days. Generally visibility is moderate in the northern and western parts of the Arabian Sea, but poor near the coast as a result of either precipitation or dust haze. At the head of the Bay of Bengal and in the Ganges Delta precipitation is associated with *monsoon depressions* which subsequently move westwards across India. Over northern and central India and in the foothills of the Himalayas rain will depend upon the position of a band of low pressure termed the *monsoon trough*.

Intermonsoon Periods

The two intermonsoon periods of April and May, and October are associated with the progressive shift of the ITCZ northwards and southwards respectively. Conditions in the ITCZ will range from calm or light winds and generally fair or fine, to squalls and cumuliform cloud with heavy rain and thunderstorms. During the months of May, June, October and November, the sea surface temperature in the Arabian Sea reaches the critical value of 27°C, and upper tropospheric conditions are more favourable for the development of tropical cyclones. In the Bay of Bengal the cyclone season normally achieves its peak in October and November, with storm surges and heavy rainfall in the Bay and along adjacent coasts.

Along the western coastline of India, onshore winds may occur during April and May. These are local sea breezes developing in response to the rapid increase of land surface temperature compared with that of the sea. In the Arabian Sea west of 55°E and within 5° to 10° of the Equator south-westerly winds occur during this period, whose strength increase to Force 7 at times. These conditions reflect the early stages in the development of the *cross-equatorial airflow*, its source being in the South Indian Ocean. The wind direction changes from south-east to south-west as it moves from the southern into the northern hemisphere. The range of latitude, over which cross-equatorial airflow is present, increases progressively to a maximum at the height of the South-West Monsoon. The feature is most pronounced over a span of 15° west of 55°E, with a core at a height of 1 to 1.5 km, and where at times the wind speed may exceed 85 knots.

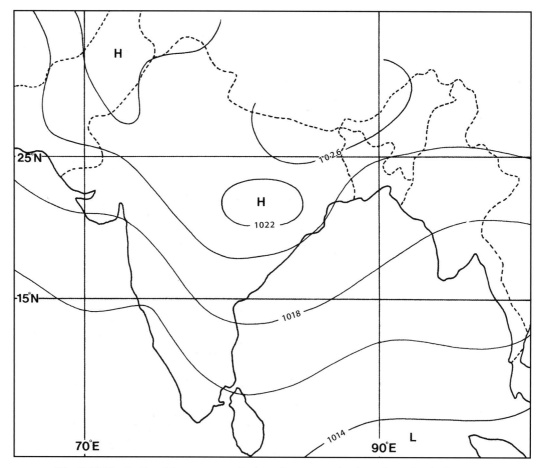

Fig. 9.22 North-East Monsoon – typical surface synoptic chart for a day in January.

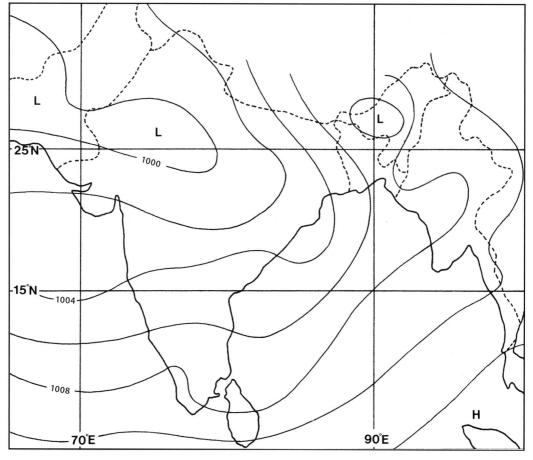

Fig. 9.23 South-West Monsoon – typical surface synoptic chart for a day in July.

North Pacific Monsoon

The annual conditions occurring in the North Indian Ocean are also experienced further east in the China and Yellow Seas, and in the western extremities of the North Pacific south of 30°N.

The North-East Monsoon first becomes evident in the northerly part of the area in September, eventually affecting the entire area north of the Equator by November (Figs. 9.20(1) and 9.24). Although referred to as the North-East Monsoon, the wind direction varies, particularly with increasing latitude. Wind strength is generally Force 6 in the Taiwan Straits and Force 5 in the China Sea, except in the area south of 10°N where Force 4 is common. Near the Equator winds are northerly and light. North of 18°N, periods of overcast skies with light rain or drizzle alternating with broken cloud cover occur between January and April. South of 18°N the cloud cover is generally four-eighths, increasing towards the Equator, with an increasing frequency of showers. These conditions persist until April when the prevailing wind direction becomes less steady and winds with a southerly component occur more frequently. This change marks the beginning of the South-West Monsoon which reaches its peak in July (Fig. 9.21).

At the height of the South-West Monsoon, wind direction varies between south-east and south-west in the area lying west of 140° E and south of 40° N. The core of the monsoon is in the South China Sea where winds of Force 3 to 4 occur. Elsewhere winds average Force 3, being lighter compared to those of the South-West Monsoon in the North Indian Ocean. Cloud cover averages four-eighths with occasional showers, except in windward coastal areas where it is greater and the rainfall heavier.

MONSOON TYPE WEATHER

In certain other low latitude areas there are seasonal changes in wind direction accompanied by a change in the amount of rainfall. Such areas do not strictly speaking have monsoons, but monsoon-type weather conditions.

Northern Australia

The movement of the ITCZ south (Fig. 9.17) contributes to the formation of the North-West monsoon along the northern coastline of Australia and the adjacent sea areas, which prevails between November and March. The flow of air is cross-equatorial from the northern hemisphere (Figs. 9.1, 9.20(1) and 9.24).

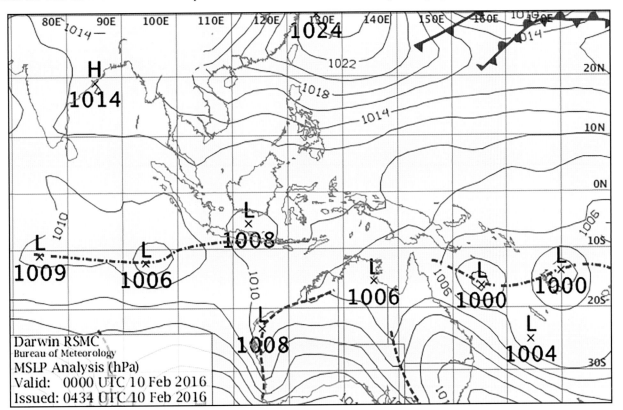

Fig. 9.24 Northern Australia Monsoon – February 10th 2016 0000 UTC

The wind direction is between west and north and the speed seldom achieving Force 7, with associated cloudy conditions and heavy rain showers. On surface charts the focal point of the flow is identified as the *monsoon trough*, the dot dash line on Figs. 9.1 and 9.24. In contrast, during the summer the South-East Trades prevail over the region (Figs. 9.19 and 9.21(1)).

Gulf of Guinea

During summer, the Gulf of Guinea and mainland Africa to the immediate north, experience south-west winds, with cloud development and precipitation (Fig 9.21(1)). This phenomenon is termed the *South-West Monsoon* and, as with Northern Australia, the focal point can be identified as the *monsoon trough*. At the beginning and the end of the South-West Monsoon, near the coast, violent thunderstorms with severe squalls moving from east to west of fairly short duration may occur. Locally these are known as *Tornadoes*.

During the remainder of the year the area has a dry season, with the wind in the Gulf of Guinea being between south and west, and the remaining area experiencing winds generally between north and east (Fig. 9.20(1)). Between November and February an easterly wind called the *Harmattan*, with a low relative humidity value transports dust and sand from the Sahara Desert. The dust and sand may be carried several hundred miles offshore generating poor visibility. At other times visibility is moderate as a result of haze.

CHAPTER 10

ORGANIZATION AND OPERATION OF
METEOROLOGICAL SERVICES

INTRODUCTION

A forecast is a statement of the anticipated meteorological conditions for either an area, or fixed location, or along a route or routes for a specified period. The term was introduced by Admiral Fitzroy, the first head of the Meteorological Office in London on its establishment in 1854. In the previous year the first international meteorological conference had been held in Brussels to discuss maritime meteorology, and in particular the system of observation at sea. In 1854, Admiral Fitzroy invited vessels to observe conditions and report their findings to his office, where climatological records were to be compiled. It was appreciated at that time that an adequate and efficient network was essential for the production of accurate records and forecasts.

THE WORLD METEOROLOGICAL ORGANIZATION

At the present time, meteorological operations are worldwide under the guidance of the World Meteorological Organization (WMO), a technical agency of the United Nations formed in 1951 and based in Geneva.

The aims of the organization are to:

1. Assist in the establishment of networks of meteorological observing stations by encouraging worldwide cooperation.

2. Assist in the development of centres to provide meteorological observations.

3. Ensure the rapid exchange of data.

4. Further the application of meteorology to human activity (e.g. shipping).

5. Encourage research and training in meteorology.

In order to achieve these aims, WMO has established a number of commissions, each concerned with a specific field. One of these is the Joint WMO/IOC Technical Commission for Oceanography and Marine Meteorology (JCOMM). As the field is large, the Commission has three distinct programme areas for observations, data management and services, and forecasting systems.

LAND OBSERVING NETWORK

One aim of WMO is that each nation should have a National Meteorological Centre (NMC), often referred to as the Central Forecasting Office (CFO) (the Meteorological Office at Exeter is the UK NMC). Linked with the NMC, usually through a number of subordinate centres, are the land observing stations, which collect surface data (Fig. 10.1). The range of meteorological elements and the frequency of observation will vary depending upon whether the observing station is classified as synoptic, auxiliary, climatological, agrometeorological or simply a health resort. Automatic Weather Stations (AWS), which are particularly valuable in inhospitable environments, are used extensively throughout the network. There are also a number of upper air stations which provide the NMC with data from upper air soundings. These stations record a full range of observations (atmospheric pressure, air temperature, humidity, wind direction and speed) daily at 0000 and 1200 UTC, but at 0600 and 1800 UTC only wind direction and speed are monitored.

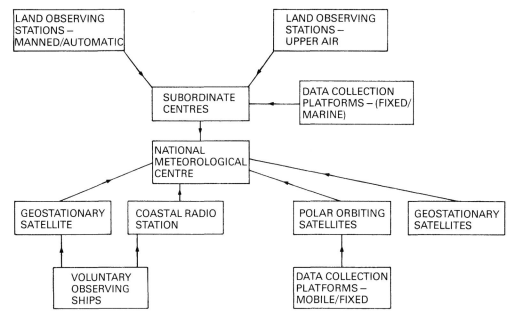

Fig. 10.1 Observing network.

SEA OBSERVING NETWORK

The national meteorological centre may also receive observations from different sources at sea (Fig. 10.1), some of which are mobile units and others are in fixed positions (Fig. 10.2).

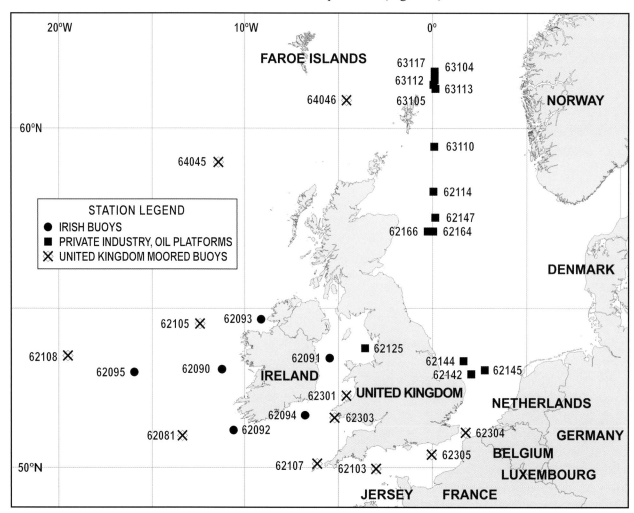

Fig. 10.2 Data buoys and platforms.

Voluntary Observing Ships

In 1853 an International Meteorological Conference held in Brussels called for voluntary surface observations at sea. The present Voluntary Observing Ships (VOS) scheme has developed from this and is now under the direction of the Ship Observations Team (SOT) of JCOMM. The scheme classifies voluntary observing ships as selected, supplementary or auxiliary mobile sea stations. Since 2009 it has also included the VOS Climate Fleet (VOSClim). Each category in the scheme is subdivided to identify vessels equipped with an Automatic Weather Station (AWS). Table 10.1 is an extract of the statistics for voluntary observing ships.

			Observing ship		
Year	*Selected*	*Supplementary*	*Auxiliary*	*VOSClim*	*Total*
1995	4124	1332	1270		6726
2000	4185	1299	1482		6966
2005	3482	908	888		5278
2010 (AWS)	2772* (109)	521* (48)	239	183* (47)	3715* (204)
2015 (AWS)	1649* (60)	630* (78)	268* (3)	498* (116)	3045* (257)

∗ Figures for AWS are included

Table 10.1 Voluntary observing ships – International.

Ships are recruited by the National Meteorological Centres of some 25 countries which have signed the International Convention for Safety of Life at Sea (SOLAS). To develop and maintain an effective observing fleet, each centre has a dedicated department (e.g. The Observations – Marine Networks department of the UK Meteorological Office), which includes Port Meteorological Officers (PMO). These officers recruit ships, which may or may not be on the country's registry, into one of the categories in the scheme depending on the vessel's trading pattern, instrument siting and the availability of the officers who will be responsible for recording and transmitting data.

Port Meteorological Officers are also responsible for the supply of instruments, which may include an AWS, instructional material, and electronic or paper meteorological logbooks. Auxiliary vessels are not provided with instruments but must have an approved simple aneroid barometer and a mercury thermometer. The PMO acts as a link through visits to the recruited vessel to check instruments, stationery, to discuss problems and collect completed logs for forwarding to the National Meteorological Centre. Table 10.2 shows the range of instrumental and visual data generally observed by each class of vessel. Visual observations may not be monitored where an AWS has been provided.

The advised synoptic hours for observations are 0000, 0600, 1200, and 1800 UTC daily, subject to ship operations. For a selected AWS equipped vessel, observations are normally monitored every three hours, but may be even more frequent. All vessels are requested to increase the frequency of observations when encountering weather conditions not forecasted. Timely transmission of observations to a meteorological centre is important. For a VOSClim vessel, however, the extra delayed-mode groups are entered in an electronic logbook for retrieval by a PMO at a later date.

As ships' meteorological observations contribute to the production of weather forecasts, the VOS Panel (VOSP), a sub-panel of the Ship Observations Team, has a critical role. Included in the remit of VOSP is the development of activities which will increase the recruitment of ships and the introduction of pilot projects.

The Automated Ship Aerological Programme(ASAP) begun in the 1980s, is overseen by the Ship Observations Team. A number of meteorological centres supply radiosonde equipment to ships with regular trading routes in the North Atlantic and North West Pacific Oceans. Ships conduct upper air soundings which are transmitted to the centres, (see Chapter 3 – Environmental Lapse Rate). Through the Worldwide Recurring ASAP Project (WRAP) the collection of data has been expanded into other ocean basins.

Class of observing ship				
Element	Selected	Supplementary	Auxiliary	VOSClim
Wind direction and speed	X	X	X	X
Atmospheric pressure	X A	X A	X	X A
Barometric tendency and characteristic	X A	O	O	X A
Air temperature	X A	X	X	X A
Dew-point temperature	X A	O	O	X A
Sea surface temperature	X	O	O	X
Waves: wind and swell	X	R	R	X
Horizontal visibility	X	X	X	X
Present and past weather	X	X	X	X
Cloud amount	X	X	X	X
Cloud type	X	X	O	X
Cloud height of base	X	X	O	X
Ship's course and speed	X	R	R	X
Sea ice and/or ice accretion	X*	X*	X*	X

XA – Observation taken by non-AWS and AWS.   O – No observation made.
X   – Observation taken by non-AWS only.   R – May be requested to observe.
X* – Observation as appropriate.

Table 10.2 Observations recorded by voluntary observing ships.

Data Buoys and Offshore Installations

Mobile and fixed data buoys are deployed worldwide. These monitor and transmit surface observations (Fig 10.2 and Fig 10.3). Data from this network is further enhanced by observations collected by offshore rigs and platforms, some of whom operate with meteorological centres, while others operate private automatic weather stations.

Fig. 10.3 Data Buoys.

METEOROLOGICAL DATA TRANSMISSION

The potential value of both surface and upper air observations in the production of a forecast depends not only on the use of standardized observing procedures, but also on the rapid transfer of data to the NMC. In order to facilitate the transfer, standard international codes of five figure groups have been developed, the total number of groups used depending upon the nature of the observations. Two examples of codes used and coded reports are shown in Fig. 10.4.

$YYGGi_w$	$99L_aL_aL_a$	$Q_cL_oL_oL_oL_o$	i_Ri_xhVV
13124	99658	10034	41497
$Nddff$	$1s_nTTT$	$2s_nT_dT_dT_d$	$4PPPP$
63423	11015	21042	40031
$5appp$	$7wwW_1W_2$	$8N_hC_LC_MC_H$	$222D_sV_s$
56002	72681	86900	22261
$0s_sT_wT_wT_w$	$2P_wP_wH_wH_w$	$3d_{w1}d_{w1}d_{w2}d_{w2}$	$4P_{w1}P_{w1}H_{w1}H_{w1}$
02073	20605	334 ⤢ ⤢	41008
$8s_wT_bT_bT_b$			
81023			

(1) Ship.

$YYGGi_w$	$IIiii$	i_Ri_xhVV	$Nddff$	$1s_nTTT$
09124	03026	42582	20000	10023
$4PPPP$	$8N_hC_LC_MC_H$			
49874	82800			

(2) Land station.

Fig. 10.4 Ship and land station code format.

There are several methods and channels through which the data may be transferred. On land, manned observing stations feed their data through a communications network via a subordinate centre to the NMC. At sea voluntary observing vessels transmit their observations by INMARSAT or e-mail. If necessary, the observations may be transmitted to a coastal radio station designated to receive weather reports. The current method of data collection from buoys employs satellites (see over) as these establish the position of the drifting buoys being interrogated.

SATELLITES

In addition to data collection and transmission, satellites are also used as remote sensing units. The history of meteorological satellites began in 1960 with the launching of a polar orbiter TIROS I (Television and Infra-red Observation Satellite). On board this satellite, *visible images,* which depend upon the reflection of solar radiation by clouds or by the surface of the earth if there are no clouds, were recorded by television camera. Currently radiometers are deployed on satellites, some sensing within the spectral range of solar radiation, from which visible images are produced (Fig. 10.5(1) and (3)). Other radiometers sense infra-red radiation which is emitted by the atmosphere and the surface, the latter depending upon cloud cover. The *infra-red image* shows varying shades of grey, normally termed a *grey scale,* which relates to the range of temperature of the masses emitting the radiation (Fig. 10.5(2) and (4)).

(1) Visible image.   (2) Infra-red image.

Fig. 10.5 Satellite images: Polar orbiter, MetOp-B – 1022 UTC 28th March 2016.

From 1975 onwards, a series of geostationary meteorological satellites became operational, enhancing the monitoring of the tropical circulation and in particular the development and progress of tropical cyclones. Dependent on the geostationary satellite, a point within the area it scans (Fig. 10.6) can be monitored anywhere between every ten to thirty minutes. Within selected areas of the disc greater frequencies of monitoring termed rapid scan, or super rapid scan are deployed, the focus being on locations of critical convective activity including tropical cyclones. Although the area sensed by a geostationary satellite is between 50°-55°N and S (Fig. 10.6), the quality of data for the mid-latitude zones is impaired since the latter lie on the periphery. The simultaneous operation of two polar orbiters with orbits at right angles to each other (Fig. 10.7) provide better coverage of the mid-latitudes and the poles, but the data is only updated every six hours. The analysis of both visible and infra-red images provides valuable information on cloud distribution and type. Geostationary cloud images are used to determine upper tropospheric winds.

A satellite may provide data which can be analysed to determine such features as vertical temperature profiles, distribution of water vapour in the upper atmosphere, sea surface temperature, and the distribution of sea ice and icebergs. Microwave radiometers fitted in satellites enable wind conditions at sea level and the distribution and type of ice on the sea surface to be determined. Geostationary satellites are also used for digital low rate information transmissions of satellite images which have been received by and processed at a surface receiving station.

(3) Visible image.

(4) Infra-red image.

Fig. 10.5 Satellite images:
Geostationary, Meteosat Seviri – 1500 UTC 21st August 2015.

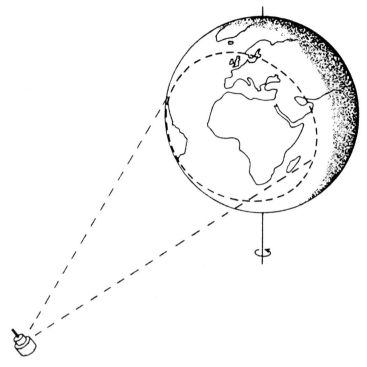

Fig. 10.6 Geostationary satellite. The illustration shows the coverage of
one satellite placed some 36000 km above the surface of the earth.

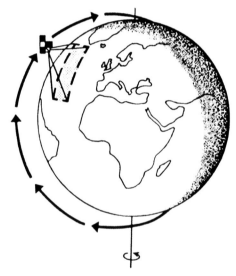

Fig. 10.7 Polar orbiting satellite. The illustration shows the path of
one satellite approximately 850 km above the surface of the earth.

GLOBAL TELECOMMUNICATIONS SYSTEM

The degree of effectiveness of all National Meteorological Centres depends not only on the efficiency of collecting and processing data from their own observing networks, but also on data from other global sources.

In 1968, World Weather Watch (WWW) was established to encourage the rapid and efficient exchange of data. An important component of WWW is the Global Telecommunications System (GTS) (Fig. 10.8), a network of three primary centres, Washington, Moscow and Melbourne, and regional hubs for the transmission of observations and processed data. The system continues to develop within the framework of the WMO Information System (WIS). The global infrastructure of WIS aims to enhance further the efficiency of the collection and dissemination of data and products, and improve access to, discovery and retrieval of weather, climate and water data by any WMO programme.

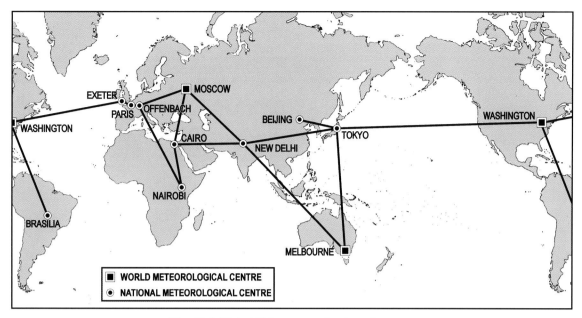

Fig. 10.8 Global Telecommunications Systems.

DATA ANALYSIS

Analysis of surface and upper air observations produces synoptic charts which are a necessary preliminary to forecasting. Observations received at an NMC are fed into a computer system which checks them for any inaccuracies, a large volume of data being processed in a relatively short period. The computer system then presents the data in the form required by the forecaster.

Surface Synoptic Charts

Surface synoptic reports are plotted by the computer using the station model format (Figs. 10.9 and 10.10). The plots are then analysed in order to construct a *surface synoptic chart,* also termed a *surface analysis.* Isobars are drawn to establish the distribution of atmospheric pressure at mean sea level, and hence the major pressure systems existing at the synoptic hour (Fig. 10.11). With the aid of satellite images (Fig. 10.5) fronts are then drawn on the chart (Fig. 10.12). The resulting surface analysis is then compared with the previous synoptic chart to confirm its feasibility.

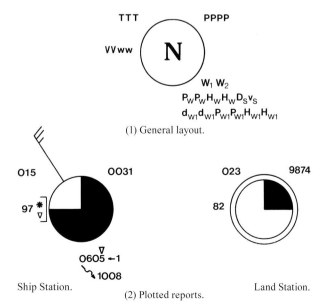

Fig. 10.9 Coded observations plotted in station model format.

(Consult Ships' Code and Decode Book Met 0.509 for symbols used for plotting).

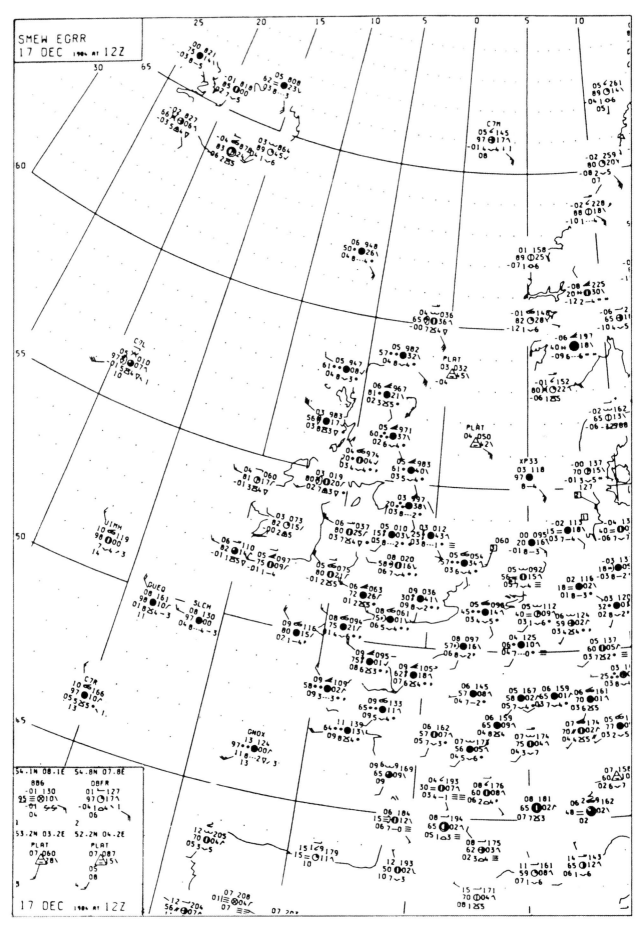

Fig. 10.10 Computer plot of surface reports.

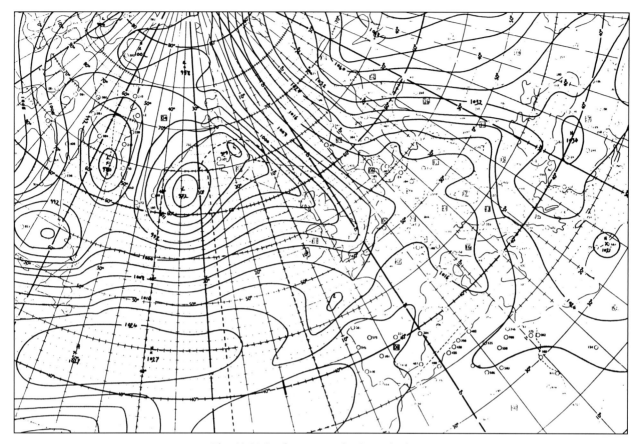

Fig. 10.11 Surface synoptic chart – isobars.

For clarity, station plots have been omitted from this chart and from Fig. 10.12.

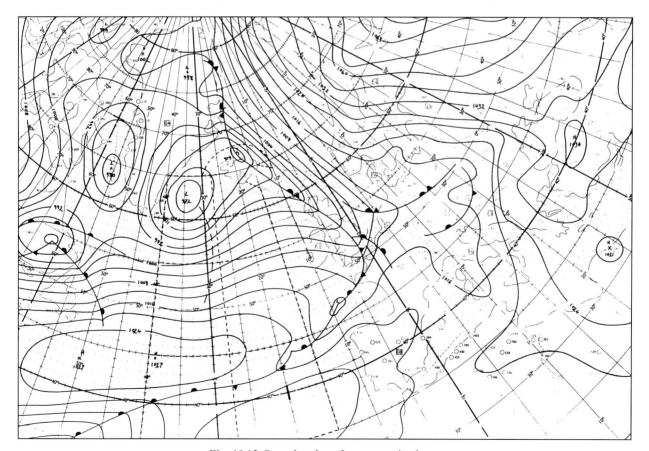

Fig. 10.12 Completed surface synoptic chart.

While the area covered by, and the frequency of surface analyses will vary from one NMC to another, charts are generally produced for the major synoptic hours of 0000, 0600, 1200 and 1800 UTC for the local regional area (e.g. British Isles and W. Europe), or for a larger area (e.g. N. Atlantic and W. Europe – Fig. 10.12). Circumpolar surface synoptic charts giving a broader view of the distribution of pressure systems are also constructed (Fig. 10.13). In addition to the above, surface synoptic charts may also be produced for intervening synoptic hours, 0300, 0900, 1500 and 2100 UTC, which generally cover the local regional area for which synoptic reports, mainly from land observing stations, are available.

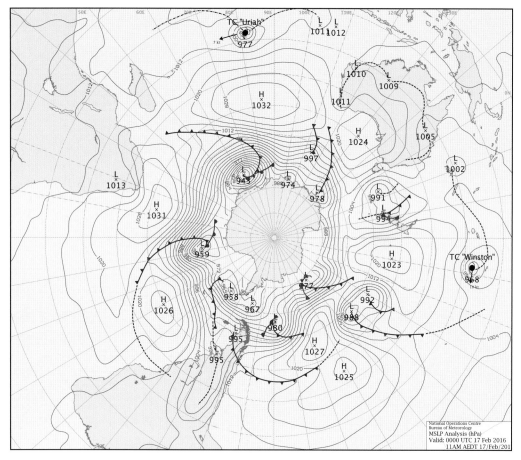

Fig. 10.13 Circumpolar surface synoptic chart.

Upper Air Charts

Upper air data collected at 0000, 0600, 1200 and 1800 UTC are transmitted to a NMC where they are checked by the computer. A number of charts can then be produced to illustrate various aspects of the upper atmosphere. One example is the *contour chart* for a given pressure level. In Fig 10.14, the contours are the blue lines with the figures in blue and black indicating the height in decametres above mean sea level at which the atmospheric pressure is 500 hPa. The brown lines are isotherms, indicating the air temperature at 500 hPa in degrees Celsius. The chart is a topographic representation of the 500 hPa level which shows a wave pattern whose amplitude and wavelength can be determined. Similar charts for other pressure levels may also be constructed, and the characteristics of jet streams determined from plotted wind data.

Upper air data is also used to compile *thickness charts,* the term referring to the vertical separation between two selected pressure levels. Fig. 10.15 illustrates a common example, the 1000 – 500 hPa thickness chart, where the brown lines known as *thickness lines* show the vertical separation in decametres. Thus the chart indicates the distribution of warm air relative to cold air, as the vertical separation is dependent upon the mean temperature of the air column between the two levels. In Fig. 10.15 the blue lines are isobars showing the distribution of pressure at mean sea level.

Both the contour and thickness charts illustrate the background details which play a critical role in the existence of the features shown on surface synoptic charts, and thus provide important and valuable information in the production of a forecast.

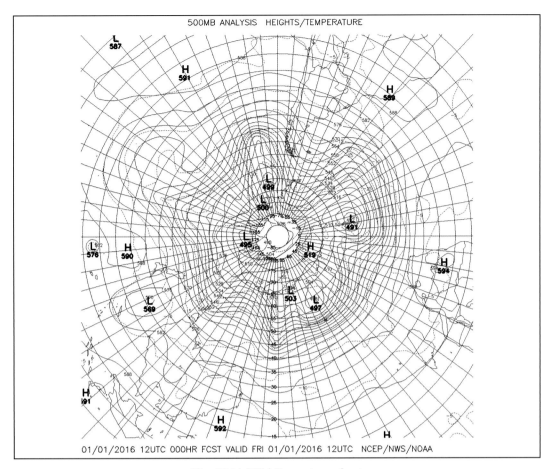

Fig. 10.14 500 hPa contour chart.

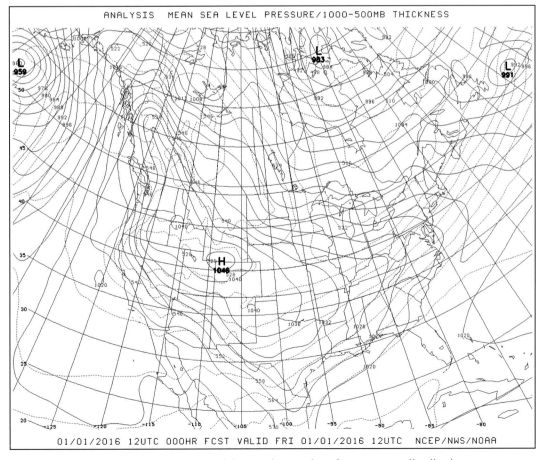

Fig. 10.15 1000–500 hPa thickness chart and surface pressure distribution.

Nephanalysis

Nephanalysis is the interpretation of visible and infra-red satellite images ("nephos" is the Greek word for cloud). Charts can be produced which identify cloud types using the basic categories of stratiform, cumuliform and cirriform. They also indicate the amount of cloud present and, in some cases, an estimate of base height. Major vortices (cyclones and anticyclones) can be identified. The characteristics of cirriform cloud may assist in locating the position of a jet stream and indicate the existence of clear air turbulence (CAT).

FORECASTING TECHNIQUES

Numerical Models

The production of forecasts involves the use of computers and numerical models, which are a series of equations representing the existing state of the atmosphere and its subsequent evolution over a given period. L. F. Richardson, a British meteorologist, devised and used the first numerical model in 1922 to produce a forecast for the area covering the British Isles and Europe. Apart from imperfections in the design of the model, which produced an excessively large pressure change for the forecast period (145 hPa in 6 hours), there was the major disadvantage of the time involved in the production of the forecast. The introduction of the electronic computer in the 1950s overcame this problem and numerical models are now part of the forecasting process.

In a numerical model the characteristics of the atmosphere are defined for a number of levels, each level being subdivided by a grid system covering a particular geographical area. The early UK model used three levels with a separation of 300 km between grid points. Greater computer power has permitted an increase in the number of levels, a decrease in distance between grid points, and an expansion of the geographical area. These refinements have improved the accuracy of forecasts. The UK Meteorological Office currently uses a global model, which has a coarse mesh. Simultaneously models covering the North Atlantic and Western Europe, and the United Kingdom are used to provide more detailed forecasts for this area.

The operation of a model depends upon an accurate input of data to determine the initial state of the atmosphere. Data from upper air soundings and analysed satellite data, particularly temperature profiles, are fed into the computer. The computer then calculates a number of characteristics of the atmosphere for each grid point at pre-determined time steps (e.g. 10 minutes) over the complete period of the forecast (e.g. 24, 48, 72, 96 and 120 hours). In the development and operation of the global model, the UK office has taken into account upper air soundings which are synoptic, and satellite data which may well be asynoptic. The data produced by the model is compared with the synoptic data for compatibility, and the latter is then incorporated into the forecasting process.

The operation of a numerical model also depends upon the input of data which includes the radiation budget of the atmosphere, and the nature of the earth's surface which affects the flux of water vapour to the atmosphere. However, present models are limited by the techniques involved and the availability of data.

Prognostic Weather Charts

The computer generated forecast data is used to produce prognostic (forecast) charts for the surface (Fig. 10.16), and a number of pressure levels in the upper atmosphere. Fronts are then drawn on the surface prognostic chart with the aid of the computer output of forecast precipitation at the surface. The final surface prognostic chart is the outcome of the interpretation of the forecast data generated by the computer, the comparison of the prognostic chart with previous synoptic charts, and the forecaster's assessment of the prognostic situation. Thus every prognostic chart is the combined production of computer and human skill, the man-machine mix.

The period ahead for which forecasts have been produced has gradually been extended. The forecast by L. F. Richardson was for a very short period compared with the 24 to 36 hour forecasts of the early operational models. Currently operational forecasts are produced for periods in excess of 5 days ahead, and work continues on improving the models.

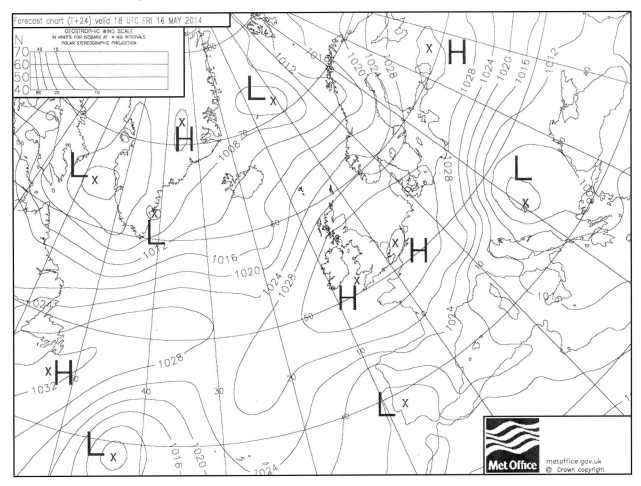

Fig. 10.16 Computer output for a surface prognostic chart.

Wave Prognostic Charts

The development of numerical models to forecast future atmospheric conditions produced a source of data which could be used to forecast wind waves and swell. Simultaneously with the demand for such information by the offshore and shipping industries, models have been developed to forecast wave fields.

This development has to a degree followed similar lines to those of atmospheric models, becoming more complex as further variables were introduced, and employing fine mesh as well as coarse mesh grids. Forecast data from the atmospheric numerical model is used to predict wind conditions over the area of the grid. Thus the probable wind direction and speed, duration and fetch are fed into the wave model. The prediction is that of the growth or decay of the wave field, which incorporates the interaction between the component of the wave field and the effects of swell and depth of water.

Predictions from wave models used by meteorological services usually include the wave period and *significant wave height,* which is defined as the mean height of the highest one third of observed waves, or those forecasted by the model. The output of the model may be total sea state, wind waves or swell, and may include wave period. The range of data presented on wave prognostic charts can vary, and in certain cases prognostic wind data may also be included (Fig 10.17(1) and (2)).

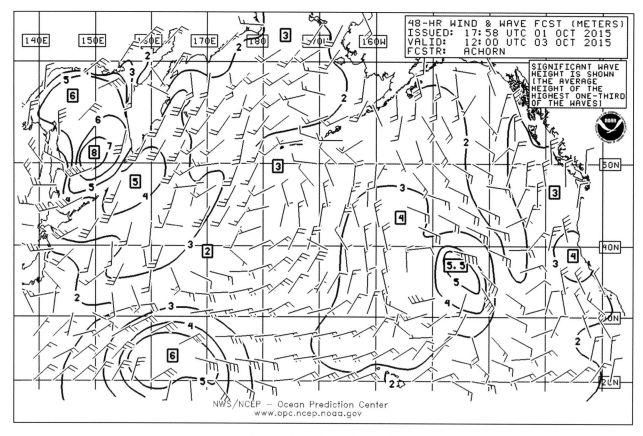

(1) 48 hour wind and wave forecast chart.

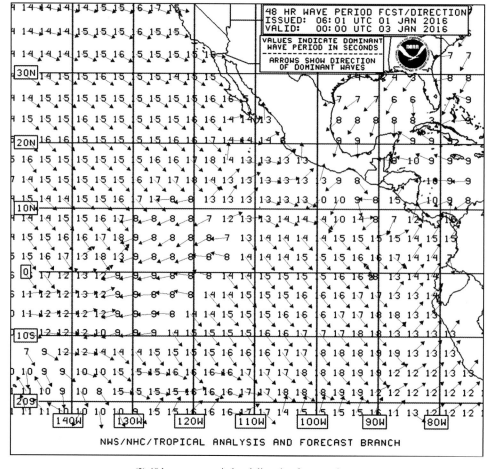

(2) 48 hour wave period and direction forecast chart.

Fig. 10.17 Wave prognostic charts.

SHIP ROUTEING SERVICES

The introduction of numerical models for forecasts led to the development of weather routeing services for ships. The general techniques were developed in the USA in the 1950s, and the service was initially offered to US military vessels, being extended to US merchant shipping in 1958. During the 1960s services offered by the Netherland Meteorological Institute at de Bilt, Seewetteramt in Hamburg, and the UK Meteorological Office (METROUTE) were introduced, whilst Ocean Routes was formed as a private US company to offer a world-wide service.

In the early days of weather routeing the provision of the service was shore based, whereby the expertise of both meteorologists and mariners employed by the organizations determined and advised a vessel of a route to follow, usually with the objective of achieving the least time on a passage. As the emphasis was on the individual nature of the service being provided, its effectiveness was dependent on the routeing organization establishing a dialogue with the master or owner obtaining information about the ship, cargo and relevant operational data for example schedules demanded by the charter, and limitations arising from insurance clauses. The information collected included ship's draught, trim, stability, speed made good relative to sea conditions (wind waves, swell, and current) which was extracted from the ship's deck log, or, if not available, obtained from a similar type of ship. The data collected was analysed to produce for each state of loading of the ship a set of performance curves, related to the important characteristics of wave height and relative direction (following, beam or head seas) (Fig. 10.18), which would be used in assessing the least-time route.

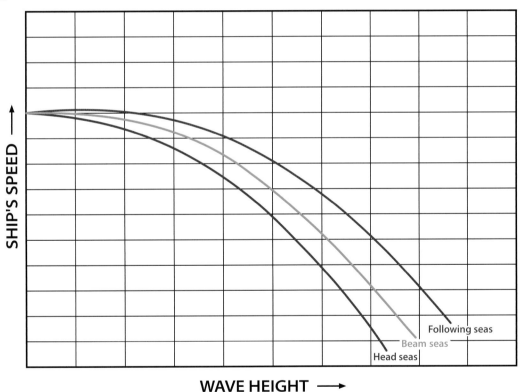

Fig. 10.18 Typical vessel performance curves.

Subsequently the range of objectives offered by routeing services expanded, for example least damage for deck cargo that needed to be protected from continuous heavy seas, or maximum fuel economy. With the shift to a wider range of objectives the service offered was renamed *ship routeing* rather than the less appropriate term of weather routeing.

Currently some organizations continue to offer the shore-based routeing services where an advised route is provided to a vessel. Another option is the preparation by an organization of forecast environmental conditions for the vessel's proposed route in the form of an optimization chart. The final decision as to the vessel's route and schedule is then made on-board by the master. Alternatively all aspects of route selection can be made on-board, whereby weather routeing software is purchased from a routeing organization, the latter providing the vessel with frequent updates of weather forecasts. Input of the forecasts and other pertinent ship's data into the software package generates an optimum and alternative routes for the vessel, with the master making the final choice. Whether ship routeing is shore based or on-board, or a combination of the two, the emphasis is on a service which is efficient and effective in the context of the crew, cargo, vessel, fuel consumption and the natural environment.

Least-time Route Selection

For the least-time objective, the initial route selected is based on the environmental conditions likely to be encountered in the area through which the ship will pass. The environmental conditions are established from synoptic and prognostic surface and upper air charts. Major pressure systems, whether depressions, anticyclones or tropical cyclones, their direction and speed of movement, development and decay, are determined. Thus the critical features of wind direction and speed, icing or reduction of visibility due to fog or heavy precipitation, are assessed together with a general appreciation of the future sea state. At this stage, synoptic and prognostic data relating to sea ice and icebergs, ocean currents from current atlases and other environmental factors are considered.

If the objective is least time, then the initial route is further evaluated by plotting either side of this route a total of five or six alternatives, 10° to 15° apart (Fig. 10.19). The speed likely to be made good along each route is determined by referring to the forecast conditions of wave height and direction from the wave prognostic chart, and relating these to the performance curves (Fig. 10.18).

Allowance is made for the set and speed of any current along the route, and the vessel's predicted distance made good over the next 12 hour run is then plotted. Thus in 12 hours the vessel could be in any one of six positions which lie on a locus also termed an *isochrone*, (Fig. 10.19). The technique is repeated for each 12 hour period ahead for which there is prognostic wave data, and a set of loci is thus plotted. A point is then selected which is nearest to the destination (Fig. 10.19), and an advised course and speed is established which would bring the vessel to that point in the least time. As the routeing organization receives more up to date wave prognostic data, so the procedure is repeated and the advised route confirmed or modified. The least time technique is thus the optimization of all environmental factors to ensure the fastest passage by the particular vessel.

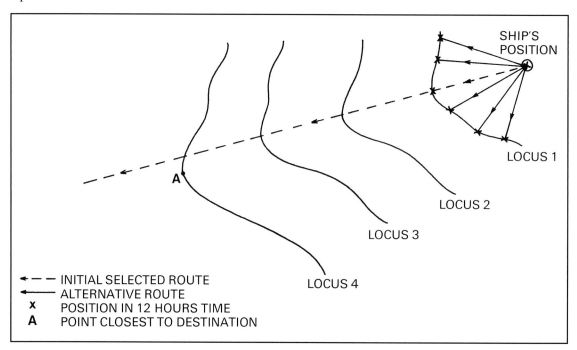

Fig. 10.19 Least-time technique.

Communications

For shore based ship routeing, communication prior to and during a voyage is vitally important. The advised route and speed is passed to the vessel before departure, and during the passage communications are maintained at intervals dictated by the reassessment of the advised route. The messages passed may also contain a summary of the synoptic and prognostic atmospheric and sea state conditions. Costs incurred for the service generally cover the fee for the advice provided and the cost of transmissions.

The progress of the vessel is monitored through the position reports made at intervals (usually 24 hours), which are recommended by the routeing organization. The reports usually include a brief summary of atmospheric and sea state conditions. In addition to these the vessel must report unforeseen circumstances (e.g. breakdown, alterations of course or speed).

Voyage Analysis

On completion of the voyage an analysis may be carried out by the routeing organization (Fig. 10.20). A synopsis of the voyage is produced in terms of meteorological and sea state conditions, and performance of the vessel over the route followed. The voyage analysis, which is made available to the vessel and the owner, often incorporates a summary of an alternative route which may have been considered.

Ship's Name: Voy   **Operators**   MetWorks — MARINE WEATHER SERVICES

Voyage Assessment: From St Eustatius **to** Antwerp

Date Time(UTC)	Steaming Time	Latitude	Longitude	Course	Distance	Speed Knots	Wind Bft	Sea Ht (m)	Wind Bft	Sea Ht (m)	Current Factor	Pformce (%)	Weather Distance
14/12/2015 10:00	Dep. Time	St Eustatius											
14/12/2015 10:00		17-33 n	063-00 w				N/A	N/A					
15/12/2015 16:00	01:06:00	21-00 n	058-00 w *	054	323	10.77	N/A	N/A	SW4	NW2.5	-0.4	98.9	4.0
16/12/2015 16:00	01:00:00	23-00 n	055-00 w*	054	235	9.79	N/A	N/A	NNW5	NW3.2	-0.3	95.8	12.1
17/12/2015 16:00	01:00:00	25-18 n	051-45 w	053	226	9.42	NNE 5-6	3-4	NNE 5	NW3.1	-0.3	96.4	10.4
18/12/2015 15:00	00:23:00	27-00 n	049-15 w	053	170	7.39	NE5-6	3-4	N5	N3.3	-0.3	94.6	14.9
19/12/2015 14:00	00:23:00	28-08 n	046-07 w	068	181	7.87	N 5-6	3-4	NW 5	N3.1	-0.3	94.8	14.4
20/12/2015 14:00	01:00:00	29-50 n	042-03 w	065	238	9.92	NNE 5	2-4	NW4	NNW 3.2	-0.4	95.8	12.1
21/12/2015 14:00	01:00:00	31-39n	037-45 w	064	248	10.33	NW5	2-4	WNW4	NW3.0	-0.4	96.8	9.2
22/12/2015 14:00	01:00:00	33-28 n	033-23 w	064	247	10.29	SW5	2-4	SSW 5	WNW 3.5	-0.4	94.3	16.4
23/12/2015 13:00	00:23:00	35-15 n	028-57 w	064	245	10.65	SSW 6-7	3-4	SSW 7	W3.4	-0.3	94.6	14.9
24/12/2015 13:00	01:00:00	37-14 n	024-24 w	062	251	10.46	WSW 5-6	3-4	SW 5	W 4.4	-0.3	91.2	25.3
25/12/2015 13:00	01:00:00	40-09 n	020-36 W	046	250	10.42	W4	4	SSW 5	W4.5	-0.4	90.6	27.1
26/12/2015 13:00	01:00:00	43-04 n	016-42 w	045	248	10.33	S7-8	4-5	SSW 7	WNW 4.1	-0.3	91.5	24.5
27/12/2015 13:00	01:00:00	45-58 n	012-34 w	046	248	10.33	S5-6	4-5	SSW 6	WNW 4.5	-0.3	90.2	28.2
28/12/2015 15:30	01:02:30	48-45 n	006-55 w	054	284	10.72	N/A	N/A	SW4	WNW 4.2	0.3	91.3	27.7
	Arr. Time												

* As no message received from vessel this position and calculations based on a DR

Charterer's Speed = 12

Voyage Summary			
Calculated Distance		3394.0	N.Miles
Voyage Time	14:05:30	341.5	Hours
Average Speed		9.94	Knots

Performance Data			
Weather Distance	241.1		N.Miles
Current Distance	99.2		N.Miles
Performance Speed	10.93		Knots

Fig. 10.20 Voyage analysis.

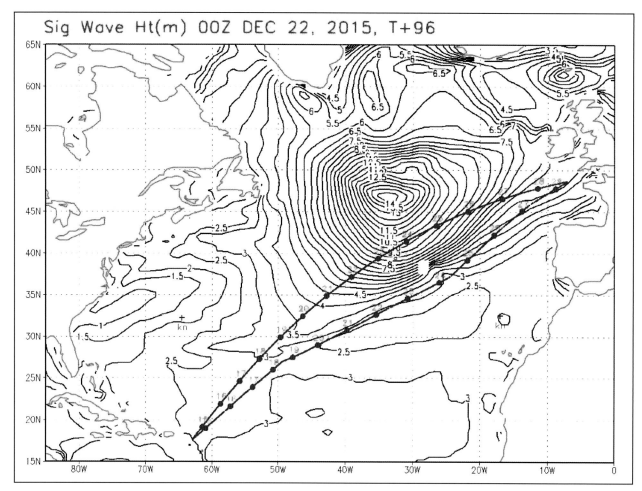

Fig. 10.21 Voyage analysis – effect of significant wave height.

Various approaches can be used to assess the benefits of routeing, one being comparison of the overall expenditure of the vessel prior to and after utilisation of a routeing service. Another demonstration of the benefits of ship routeing is an estimate of the time saved by the vessel on the recommended route compared with the transit time if it had taken an alternative route (Fig. 10.21).

CHAPTER 11

FORECASTING SOURCES

While the seafarer can do nothing to control the atmospheric and sea conditions affecting his vessel, he can indirectly influence the circumstances arising from them through careful evaluation and application of forecast data both in pre-passage planning and on passage. The success or otherwise of the evaluation and application depends upon the amount and quality of data available.

SINGLE OBSERVER FORECASTING

Observations and Analysis

Whether on passage or in port the seafarer can directly monitor his environment to a greater or lesser degree, depending upon the range of facilities and the time available. By analysing a comprehensive series of observations (Table 11.1) trends can be established, from which the recent history of the atmosphere and sea conditions can be determined. As an example, pressure tendency, not necessarily restricted to the standard three hours preceding the time of observation, can be assessed from the barogram (Chapter 2), or established to the nearest hectopascal by frequent reading of the simple aneroid barometer. Trends can also be assessed of the visibility range, cloud amount, and sequence of cloud type. The value of visibility trends are enhanced when related to the presence and intensity of precipitation, mist, fog, haze or the sea state. The tendency of the wind to back or veer, and the speed to increase or decrease can be determined, and the characteristics of swell are also worth noting (Chapter 9).

Date	GMT	Atmospheric pressure (hPa)	Air temp. °C	Wind		Cloud		Comments
				Direction	Force	Cover	Type	
13th	2000	991.5	7.0	S	5	8/8	Cs	Moderate sea.
14th	0400	985.3	7.3	SSE	6	8/8	Ns	Rain for past two hours, rough sea.
14th	0730	979.4	8.9	SE	6-7	8/8	Ns	Heavy rain, rough sea.

Table 11.1 Onboard observations.

Forecast

The analysis of the data in Table 11.1 indicates the advance of a frontal depression, and the bearing of the centre can be determined by applying Buys Ballot's Law. Thus conditions likely to be encountered will be those associated with the passage of such a depression (Chapter 8). Once such a short term forecast is made, the monitoring process should be continued to verify or modify the forecast at frequent intervals, as the conclusions reached will be important for a safe passage.

The above procedure has inherent limitations, as it assumes that the particular pressure system inferred from the observations will result in the same meteorological conditions on every occasion. Variations will occur and the knowledge of the seafarer, who may have great experience of the area, will be valuable in the analysis and forecast procedures.

ISSUED METEOROLOGICAL DATA

A wealth of meteorological data is available to the seafarer both at sea and in port. In accordance with WMO recommendations National Meteorological Centres issue data in return for the observations received from voluntary observing ships. The radio weather services available worldwide to the seafarer are documented in the Admiralty List of Radio Signals, Volume 3 (ALRS Vol. 3). ALRS Vol. 4 lists meteorological observation stations. Other similar publications where such details may be found are listed in Table 11.2.

Bollettini Meteorologici per la Navigazione Marittima (Italy).

ITU List of Radiodetermination and Special Service Stations (Italy).

Lista de Radioajudas (Portugal).

Nautischer Funkdienst, Band III (Germany).

New Zealand Nautical Almanac (New Zealand).

Radioayudas a la Navegacion (Argentina).

Radioayudas a la Navegacion en la Costa de Chile (Chile).

Radiosignaux Meteorologiques (France).

Radio Aids to Marine Navigation (Canada).

WMO Publication No. 9.

Worldwide Marine Weather Broadcasts (United States).

Table 11.2 Publications listing radio weather services.

The areas of responsibility of each country for weather and sea bulletins are shown in Appendix 2.

Storm Warnings

Forecasters in an NMC, having established from prognostic data the possibility of high winds over sea areas, will compile a "Storm Warning". The anticipated wind force and direction for the sea areas likely to be affected are stated, and, if the warning is due to a tropical cyclone, the position of its centre, and its past and future movement are included. Radio stations listed in ALRS Vol. 3 and ALRS Vol. 5 with the sub-heading "Storm Warnings" transmit these messages on their working frequencies, the modes of transmission being either R/T or W/T or both. The transmission is normally made at the end of the next silence period after receipt of the message. In certain cases the message is repeated, the interval between broadcasts and the overall period in which transmissions are made varying from one station to another.

The sub-heading "Storm Warnings" is general rather than specific, since the forecast wind speed may vary from Force 6 (Strong Breeze) to Force 12 (Hurricane). The message is usually issued in the language of the country of origin, and WMO recommends that it is also issued in English.

In certain parts of the world the national broadcasting service of a country is used as a channel for transmitting storm warnings (ALRS Vol. 3). The transmissions are made as soon as possible after the receipt of the warning from the NMC (normally at the first available programme break), and are repeated at convenient breaks so long as the warning remains in force.

If the seafarer is to gain the maximum benefits of the storm warning service, then a policy of regular radio watch is essential. However, delays which may occur between the original issue of a warning and its receipt on board can seriously reduce its value, which is critical in the case of tropical or temperate zone low pressure systems since these can develop and deepen rapidly. The information in a warning is intended to establish concisely the forecast conditions for the sea area concerned, but it must be interpreted intelligently using onboard observations and other data, as it may only be applicable to part of the area.

Weather Bulletins

National Meteorological Centres also issue data in the form of weather bulletins for vessels at sea. As with storm warnings the channels used are the radio stations, and the operating frequencies, times, and method (W/T) of transmission are listed in ALRS Vol. 3 under the sub-heading "Weather Messages". ALRS Vol. 5 should also be consulted. Whilst the nature of bulletins varies around the world, WMO does recommend a certain range and order of presentation of data.

A bulletin may consist of six parts (Table 11.3). Parts 1, 2 and 3 contain the important data, usually in the language of the country of origin, but may be repeated in English and the terminology used normally follows WMO guidelines. If Parts 4, 5 and 6 are included, they may be transmitted with Parts 1, 2, and 3, or at a later time.

Part	Title	Contents
1	Storm Warnings	Sea areas affected identified, forecast wind conditions, and the location, movement, development or decay of the related pressure centres. The statement of "No Storm Warnings" is used when there is no forecast storm condition.
2	Synopsis	Identifies the major pressure systems, their movement, development or decay, and if available, significant sea and swell conditions and ice conditions where applicable. In low latitudes this part may simply state the existence of seasonal weather, the daily condition being similar to the seasonal pattern and the rate of change minimal.
3	Forecast	A forecast statement for the period stated in the bulletin for each of the sub-sections for which the NMC is responsible. The statement includes the anticipated wind, visibility, weather and wave conditions, ice accretion and ice conditions where applicable. An outlook for a period beyond twenty-four hours should also be included.
4	Analysis or prognosis	Details of type and location of pressure centres, central pressure values, types and location of fronts, and position of selected isobars. The data is in coded format using the IAC FLEET Code (an abbreviated form for marine use of the International Analysis Code (IAC)).
5	Sea station reports	Synoptic reports from vessels in a reduced code format.
6	Land station reports	Synoptic reports from coastal land stations in a reduced code format.

Table 11.3 Structure of a weather bulletin.

As an example the UK Meteorological Office issues the High Seas Weather Bulletin comprising Parts 1, 2 and 3 (Table 11.4). The sea areas in Part 3 are shown in Fig. 11.1. The bulletin, which is prepared twice daily, is transmitted via INMARSAT which is part of the Global Maritime Distress and Safety System (GMDSS). Similar bulletins are issued by various meteorological services which are responsible for their own sea areas (Appendix 2). The times of transmission of a particular bulletin via INMARSAT and the sea areas covered are listed in ALRS Vol. 5.

By using a weather bulletin in conjunction with a surface analysis received by facsimile, a seafarer can generate a prognostic situation by correlating the forecast data in Part 3 with the movement, development or decay of the pressure systems shown on the chart.

HIGH SEAS BULLETIN FOR METAREA 1

ISSUED AT 0800 UTC ON SUNDAY 27 MARCH 2016 BY THE MET OFFICE, EXETER, UNITED KINGDOM FOR THE PERIOD 0800 UTC ON SUNDAY 27 MARCH UNTIL 0800 UTC ON MONDAY 28 MARCH 2016

STORM WARNING

AT 270000UTC, LOW 58 NORTH 07 WEST 966 EXPECTED 68 NORTH 03 EAST 963 BY 280000UTC. SOUTHERLY OR SOUTHWESTERLY WINDS WILL REACH STORM FORCE 10 OR VIOLENT STORM FORCE 11 IN CENTRAL AND EASTERN NORWEGIAN BASIN UNTIL 280000UTC

GENERAL SYNOPSIS

AT 270000UTC, LOW 58 NORTH 07 WEST 966 EXPECTED 68 NORTH 03 EAST 963 BY 280000UTC. LOW 48 NORTH 36 WEST 999 EXPECTED 51 NORTH 05 WEST 977 BY SAME TIME. LOW 54 NORTH 17 WEST 975 EXPECTED 57 NORTH 07 WEST 982 BY THAT TIME. AT 270000UTC, LOW 67 NORTH 08 WEST 976 LOSING ITS IDENTITY. HIGH 68 NORTH 37 WEST 1007, SLOW-MOVING WITH LITTLE CHANGE

AREA FORECASTS FOR THE NEXT 24 HOURS

SOLE

SOUTHWEST 7 TO SEVERE GALE 9, BECOMING CYCLONIC, THEN NORTHWEST LATER, 6. VERY ROUGH OR HIGH, OCCASIONALLY VERY HIGH AT FIRST. RAIN OR THUNDERY SHOWERS. GOOD, OCCASIONALLY POOR

SHANNON

WEST 7 TO SEVERE GALE 9, DECREASING 6. HIGH OR VERY HIGH, BECOMING ROUGH OR VERY ROUGH LATER. THUNDERY SHOWERS. GOOD, OCCASIONALLY POOR

ROCKALL

WESTERLY OR SOUTHWESTERLY 6 TO GALE 8 AT FIRST IN SOUTH, OTHERWISE CYCLONIC 5. VERY ROUGH OR HIGH, BECOMING ROUGH LATER. WINTRY SHOWERS. GOOD, OCCASIONALLY POOR

BAILEY

NORTH OR NORTHWEST 5 OR 6. ROUGH OR VERY ROUGH. WINTRY SHOWERS. GOOD

FAEROES

CYCLONIC 7 TO SEVERE GALE 9, BECOMING WEST 5 OR 6, THEN CYCLONIC 5 LATER. VERY ROUGH OR HIGH, OCCASIONALLY ROUGH LATER. RAIN THEN SHOWERS. MODERATE, OCCASIONALLY POOR

SOUTHEAST ICELAND

NORTHERLY 7 TO SEVERE GALE 9, DECREASING 6 LATER. VERY ROUGH OR HIGH, OCCASIONALLY ROUGH IN WEST. WINTRY SHOWERS. MODERATE, OCCASIONALLY POOR

EAST NORTHERN SECTION

IN NORTH, NORTHEASTERLY, BECOMING CYCLONIC AT TIMES IN SOUTH, 5 TO 7, OCCASIONALLY GALE 8 AT FIRST IN FAR NORTHWEST. MODERATE OR ROUGH, OCCASIONALLY VERY ROUGH. WINTRY SHOWERS, OCCASIONAL SNOW LATER. MODERATE OR GOOD, OCCASIONALLY POOR, BUT BECOMING VERY POOR AT TIMES LATER. IN SOUTH, CYCLONIC, MAINLY NORTHWESTERLY, 4 OR 5, INCREASING 6 AT TIMES. ROUGH OR VERY ROUGH, BECOMING MODERATE OR ROUGH. WINTRY SHOWERS. GOOD, OCCASIONALLY POOR

WEST NORTHERN SECTION

IN NORTH, CYCLONIC, BECOMING NORTHEASTERLY, 5 TO 7, OCCASIONALLY GALE 8 AT FIRST, DECREASING MAINLY 4 LATER IN FAR NORTHWEST. ROUGH OR VERY ROUGH, OCCASIONALLY MODERATE LATER. OCCASIONAL SNOW OR WINTRY SHOWERS. MODERATE OR GOOD, OCCASIONALLY VERY POOR. IN SOUTH, NORTHERLY OR NORTHWESTERLY, BECOMING CYCLONIC IN EAST, 5 TO 7, OCCASIONALLY GALE 8 IN FAR SOUTHWEST. ROUGH OR VERY ROUGH. OCCASIONAL SNOW OR WINTRY SHOWERS. MODERATE OR GOOD, OCCASIONALLY VERY POOR

EAST CENTRAL SECTION

IN NORTHEAST, WESTERLY OR NORTHWESTERLY 6 TO GALE 8, DECREASING 4 OR 5. VERY ROUGH OR HIGH, OCCASIONALLY VERY HIGH AT FIRST IN EAST, BECOMING MODERATE OR ROUGH LATER. WINTRY SHOWERS. GOOD, OCCASIONALLY POOR. IN SOUTH, CYCLONIC 7 TO SEVERE GALE 9, BECOMING WESTERLY OR NORTHWESTERLY 6 TO GALE 8. VERY ROUGH OR HIGH, OCCASIONALLY VERY HIGH AT FIRST IN EAST. RAIN, THEN WINTRY SHOWERS. MODERATE OR POOR, BECOMING GOOD, OCCASIONALLY POOR. IN NORTHWEST, NORTHWESTERLY 5 OR 6, INCREASING 7 OR GALE 8 LATER IN SOUTH. ROUGH OR VERY ROUGH. WINTRY SHOWERS. GOOD, OCCASIONALLY POOR

WEST CENTRAL SECTION

NORTHWESTERLY 6 TO GALE 8. ROUGH OR VERY ROUGH, OCCASIONALLY HIGH. RAIN, THEN SNOW SHOWERS. MODERATE OR GOOD, OCCASIONALLY VERY POOR

DENMARK STRAIT

NORTHEASTERLY 6 OR 7, DECREASING 4 OR 5 LATER. ROUGH OR VERY ROUGH, OCCASIONALLY MODERATE LATER IN WEST. SNOW SHOWERS. MODERATE OR GOOD, OCCASIONALLY VERY POOR. LIGHT OR MODERATE ICING AT FIRST IN FAR EAST WITH TEMPERATURES MS03 TO MS06

NORTH ICELAND

IN WEST, NORTHEASTERLY 6 TO GALE 8, BACKING NORTHERLY 4 OR 5 LATER. VERY ROUGH OR HIGH, BECOMING ROUGH OR VERY ROUGH LATER. SNOW SHOWERS. MODERATE OR GOOD, OCCASIONALLY VERY POOR. LIGHT OR MODERATE ICING WITH TEMPERATURES MS03 TO MS06, SEVERE OR VERY SEVERE ICING LATER IN FAR NORTH WITH TEMPERATURES MS09 TO MS12. IN EAST, CYCLONIC IN FAR SOUTH, OTHERWISE NORTHEASTERLY, 6 TO GALE 8, BACKING NORTHERLY 6 OR 7 LATER. VERY ROUGH OR HIGH. OCCASIONAL SNOW. MODERATE OR GOOD, OCCASIONALLY VERY POOR. LIGHT OR MODERATE ICING WITH TEMPERATURES MS03 TO MS06 EXCEPT IN FAR SOUTH, BECOMING SEVERE OF VERY SEVERE ICING IN FAR NORTH WITH TEMPERATURES MS09 TO MS12

NORWEGIAN BASIN

IN AREA WEST OF 00 EAST, CYCLONIC 7 TO SEVERE GALE 9, BECOMING SOUTHWESTERLY 5 OR 6 LATER. VERY ROUGH OR HIGH, OCCASIONALLY ROUGH. RAIN OR SHOWERS. MODERATE OR GOOD, OCCASIONALLY POOR. IN AREA EAST OF 00 EAST, SOUTHERLY OR SOUTHWESTERLY SEVERE GALE 9 OR STORM 10, INCREASING VIOLENT STORM 11 AT TIMES, DECREASING 6 TO GALE 8 LATER. VERY ROUGH OR HIGH, BECOMING VERY HIGH FOR A TIME. RAIN OR SHOWERS. MODERATE OR GOOD, OCCASIONALLY POOR

OUTLOOK FOR FOLLOWING 24 HOURS:

GALES OR SEVERE GALES EXPECTED IN SOLE, SHANNON, BOTH NORTHERN SECTIONS, BOTH SOUTHERN SECTIONS, NORTH ICELAND AND NORWEGIAN BASIN

Table 11.4 High Seas Weather Bulletin – 27th March 2016

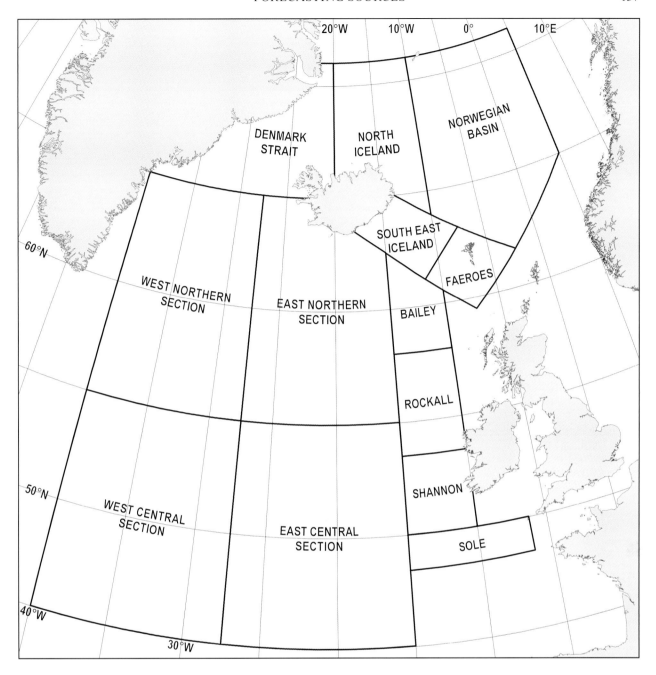

Fig. 11.1 High Seas Weather Bulletin – Forecast areas.

Maritime Forecast Code

National Meteorological Centres may issue bulletins in their own language only, and repeat the information using the Maritime Forecast Code, the coded message being prefixed by MAFOR. The area to which the forecast applies is identified by an indicator figure or geographical name. The period for which the forecast is valid and the mean forecast values for a number of meteorological elements are given in code format. The significant elements are surface wind and weather (e.g. precipitation, fog, mist, ice accretion, visibility).

Forecast data relating to sea state, air temperature and swell may also be included. In a transmission these elements may be considered more than once so describing any changes which may occur and their duration. Thus the forecast may cover a period up to 72 hours ahead, which is of value to the seafarer when related to the assessment of onboard observations.

Shipping Forecast

The UK Meteorological Office issues a shipping forecast for coastal areas, which is broadcast by the British Broadcasting Corporation (BBC) on Radio 4. The contents of the transmission include gale warnings, a general synopsis (with comments on future change), a forecast for the next 24 hours of the wind, weather and visibility for each coastal area and the latest reports from a number of coastal stations. The forecast is issued four times daily, and provides in greater detail the anticipated conditions for sea areas around the British Isles (Fig. 11.2).

For inshore waters (up to 12 miles offshore) forecasts of wind, weather and visibility are broadcast on BBC Radio 4. Synoptic reports from a selection of observing stations are included in the Radio 4 broadcasts. These forecasts are valuable both to the seafarer and to those involved in recreational activities.

THE SHIPPING FORECAST ISSUED BY THE MET OFFICE, ON BEHALF OF THE MARITIME AND COASTGUARD AGENCY, AT 1725 UTC ON SUNDAY 27 MARCH 2016 FOR THE PERIOD 1800 UTC SUNDAY 27 MARCH TO 1800 UTC MONDAY 28 MARCH 2016
THERE ARE WARNINGS OF GALES IN VIKING NORTH UTSIRE TYNE DOGGER FISHER GERMAN BIGHT HUMBER THAMES DOVER WIGHT PORTLAND PLYMOUTH BISCAY TRAFALGAR FITZROY SOLE LUNDY FASTNET IRISH SEA AND SOUTHEAST ICELAND
THE GENERAL SYNOPSIS AT MIDDAY *LOW MALIN 980 EXPECTED JUST WEST OF ROCKALL 983 BY MIDDAY TOMORROW. LOW 300 MILES WEST OF FITZROY 988 EXPECTED HUMBER 974 BY SAME TIME*
THE AREA FORECASTS FOR THE NEXT 24 HOURS ***VIKING NORTH UTSIRE*** *SOUTHERLY BACKING SOUTHEASTERLY LATER, 6 TO GALE 8. VERY ROUGH OR HIGH, BECOMING MODERATE OR ROUGH LATER. SHOWERS, RAIN LATER. GOOD, OCCASIONALLY POOR LATER* ***SOUTH UTSIRE EAST FORTIES*** *SOUTHERLY BACKING SOUTHEASTERLY LATER, 5 TO 7 PERHAPS GALE 8 LATER. ROUGH, OCCASIONALLY VERY ROUGH FOR A TIME IN NORTH. SHOWERS, RAIN LATER. GOOD, OCCASIONALLY POOR LATER* ***WEST FORTIES CROMARTY FORTH*** *SOUTHERLY 5 TO 7, BECOMING VARIABLE 4 LATER, THEN NORTHWESTERLY 5 OR 6. MODERATE OR ROUGH. SHOWERS. GOOD* ***TYNE DOGGER*** *SOUTH 5 TO 7, BECOMING CYCLONIC 6 TO GALE 8 LATER, PERHAPS SEVERE GALE 9 IN DOGGER. MODERATE OR ROUGH, OCCASIONALLY VERY ROUGH LATER IN EAST DOGGER. RAIN. GOOD, OCCASIONALLY POOR* ***FISHER*** *SOUTH 5 TO 7, BACKING SOUTHEAST 6 TO GALE 8 LATER. MODERATE OR ROUGH, BECOMING ROUGH OR VERY ROUGH LATER. SHOWERS, RAIN LATER. GOOD, OCCASIONALLY POOR LATER* ***GERMAN BIGHT HUMBER THAMES*** *SOUTH 7 TO SEVERE GALE 9, VEERING SOUTHWEST GALE 8 TO STORM 10 LATER. MODERATE OR ROUGH, BECOMING ROUGH OR VERY ROUGH LATER. RAIN. GOOD, OCCASIONALLY POOR* ***DOVER WIGHT PORTLAND*** *WEST OR SOUTHWEST 6 TO GALE 8, BACKING SOUTH GALE 8 TO STORM 10 FOR A TIME. ROUGH OR VERY ROUGH. RAIN OR THUNDERY SHOWERS. MODERATE, OCCASIONALLY POOR* ***PLYMOUTH BISCAY FITZROY*** *WEST OR SOUTHWEST 7 TO SEVERE GALE 9, OCCASIONALLY STORM 10 UNTIL LATER. VERY ROUGH OR HIGH, OCCASIONALLY VERY HIGH FOR A TIME IN BISCAY AND FITZROY. RAIN OR THUNDERY SHOWERS. MODERATE, OCCASIONALLY POOR* ***SOLE*** *CYCLONIC BECOMING WEST 7 TO SEVERE GALE 9. VERY ROUGH OR HIGH. OCCASIONAL RAIN. GOOD, OCCASIONALLY POOR* ***LUNDY FASTNET*** *CYCLONIC 6 TO GALE 8 OCCASIONALLY SEVERE GALE 9 IN LUNDY, BECOMING NORTHWEST 5 TO 7, BACKING WEST 6 TO GALE 8 LATER. VERY ROUGH OR HIGH, BECOMING ROUGH OR VERY ROUGH. OCCASIONAL RAIN. GOOD, OCCASIONALLY POOR* ***IRISH SEA*** *SOUTHWEST 6 TO GALE 8, VEERING NORTH 5 OR 6, BACKING SOUTHWEST 4 OR 5 LATER. ROUGH OR VERY ROUGH, BECOMING SLIGHT OR MODERATE LATER. SHOWERS. GOOD* ***SHANNON ROCKALL MALIN*** *WEST OR NORTHWEST 5 OR 6, OCCASIONALLY 7 IN SHANNON. ROUGH OR VERY ROUGH. SHOWERS. GOOD* ***HEBRIDES SOUTH BAILEY*** *CYCLONIC 4 OR 5, OCCASIONALLY 6 LATER. ROUGH, OCCASIONALLY VERY ROUGH FOR A TIME. SHOWERS. GOOD* ***NORTH BAILEY*** *NORTHEASTERLY 5 TO 7. ROUGH, OCCASIONALLY VERY ROUGH FOR A TIME. SHOWERS. GOOD* ***FAIR ISLE*** *CYCLONIC 5 TO 7. ROUGH OR VERY ROUGH, BECOMING MODERATE OR ROUGH. SHOWERS. GOOD* ***FAEROES SOUTHEAST ICELAND*** *NORTH OR NORTHWEST, VEERING NORTHEAST LATER, 5 TO 7 OCCASIONALLY GALE 8 AT FIRST IN SOUTHEAST ICELAND. ROUGH OR VERY ROUGH, OCCASIONALLY HIGH FOR A TIME IN SOUTHEAST ICELAND. WINTRY SHOWERS. GOOD, OCCASIONALLY POOR*

Table 11.5 Shipping forecast – 27th March 2016

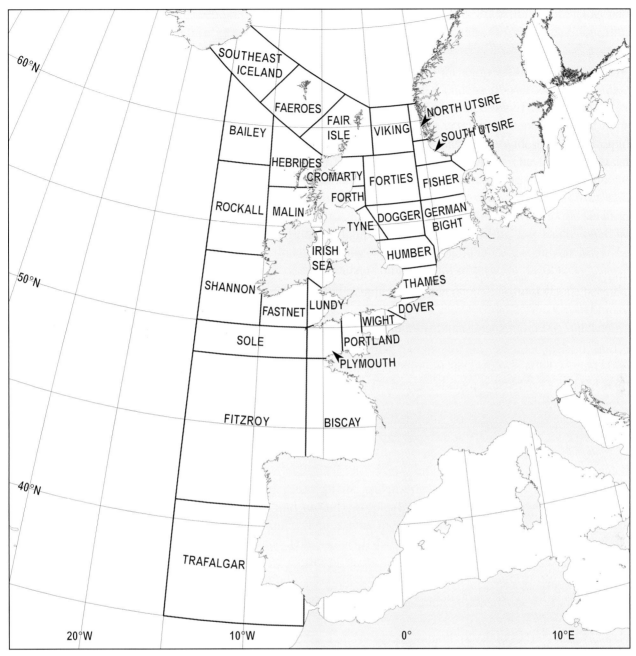

Fig. 11.2 Shipping Forecast areas.

Navigational Warnings

Meteorological warnings of conditions affecting safety of vessels (ALRS Vol. 3 and Vol. 5) may either be included in the NAVAREA warnings or precede their issue. The phenomena which may form a part or all of the broadcast may be tropical cyclones, visibility of 2 km or less, severe icing, wave heights of 8 m or more, and winds of Force 10 or above. Gale warnings, weather forecasts and ice reports are also transmitted via NAVTEX (ALRS Vol. 5).

FACSIMILE CHARTS

Equipment

Several companies manufacture and market facsimile equipment comprising a receiver and recorder designed specifically for meteorological charts (Fig. 11.3). The seafarer, by consulting ALRS Vol. 3 can identify a facsimile station which will transmit the data required. The station frequency can then be selected. When the automatic mode on the recorder is selected, the recorder is activated by the initial signal transmitted which also sets the drum speed (revolutions per minute) and *index of co-operation* (a term relating to the line scanning of the equipment). The recorder is also stopped automatically at the end of the transmission. If there is no automatic mode, the operator must set the recommended drum speed and index of co-operation. When the transmission begins the recorder is started, and the alignment of the paper checked and adjusted if necessary so that the chart is printed as a single entity. The equipment must be stopped manually at the end of the transmission. However, it should be noted that the procedure for operating facsimile equipment varies and operating manuals should always be consulted. The total time taken for a chart to be transmitted will depend upon its size.

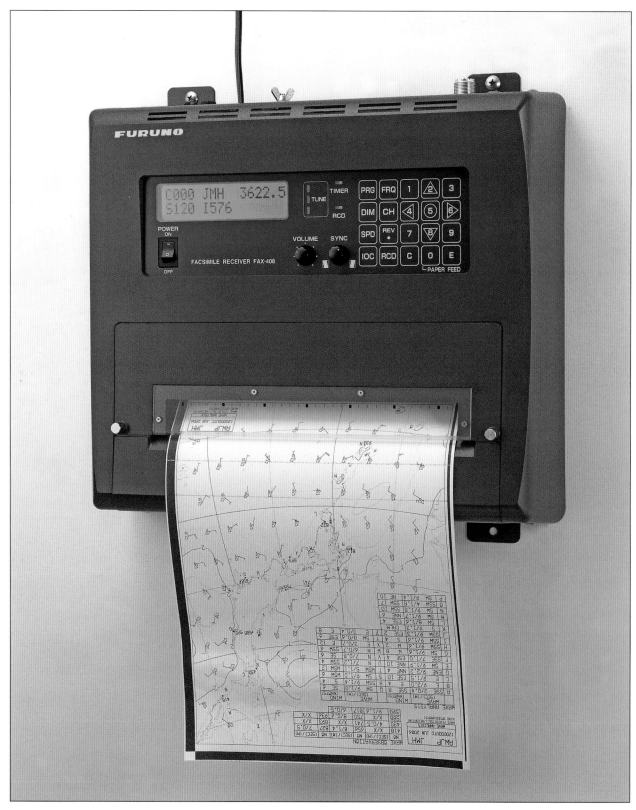

Fig. 11.3 Facsimile Receiver – Fax-408 (Furuno).

Surface Analysis

The facsimile schedules in ALRS Vol. 3 provide information on the types of charts available from meteorological centres. Of those listed, the surface analysis and surface prognostic charts are of particular value to the seafarer. The surface analysis shows the surface distribution of pressure by means of isobars and associated fronts for the synoptic hour stated (Fig. 11.6). Synoptic reports from land stations and ships may also be included using the recommended station plot format, or, if there is insufficient space, only critical elements may be plotted (e.g. wind direction and speed). Although the chart is essentially synoptic in nature, prognostic data for direction and speed of movement, and central pressure values of pressure centres may also be included. Retrospective data for pressure centres, a horizontal distance scale, and a geostrophic wind scale may also be shown. The chart projection may be noted and, for clarity and quick reference, latitude and longitude intersections are plotted at frequent intervals. If the chart is a preliminary analysis, the data shown is that of isobars which illustrate the major pressure systems.

The area covered by a facsimile surface analysis varies from one transmitting source to another, and the seafarer can check its limits from the schedule. Surface analyses from any one source may cover different areas, and while a few meteorological offices may issue one chart daily, most issue at least two, one for 0000 UTC and one for 1200 UTC. Generally the number of charts issued is greater for the immediate area of interest to the meteorological office compared with the number covering a larger horizontal area.

Prognostic Charts

Surface prognostic charts show the anticipated distribution of surface pressure by means of isobars and related fronts for a future time, which is noted in its title. Figs. 11.7 and 11.8 are examples of the range of surface prognostic charts issued by the Japan Meteorological Agency which are listed in the facsimile schedule in ALRS Vol. 3. As with the surface analysis, the issue of prognostic charts varies from one source to another. Generally the 24 hour surface prognosis is issued at least once daily, and in some cases four times daily for the synoptic hours 0000, 0600, 1200 and 1800 UTC. However, surface prognostic charts for times in excess of 24 hours may only be issued once daily. It should also be noted that, although a degree of continuity of area is maintained for the more important charts, the area covered by a 24 hour surface prognostic chart may not be the same as that of the 36 hour chart from the same source.

Facsimile prognostic data from some sources may not be presented in the form described above. Instead a surface prognostic chart may show the anticipated wind direction and speed for a number of sea areas by means of symbols, and significant weather conditions in plain language. Plain language prognostic and synoptic statements may also be issued via the facsimile system.

Other Charts

A number of other charts may be available to the seafarer through facsimile transmissions. Among them are the wave analysis and wave prognostic charts, the latter being for 24 and 48 hours (Fig. 10.17). Ice charts, whose transmission schedules are to be found in ALRS Vol. 3, may include data relating to sea ice, icebergs and sea surface temperature (Figs. 11.4 and 11.5). Certain ice charts may only show synoptic data but on others prognostic data may be included, as with those issued by the International Ice Patrol. Ice charts can usually be supplemented by bulletins from a number of sources which are listed in ALRS Vol. 3.

Satellite images or nephanalyses are issued by some meteorological centres (Chapter 10). A limited number of centres may issue radar summaries, data on surface ocean currents, and ocean thermal analyses.

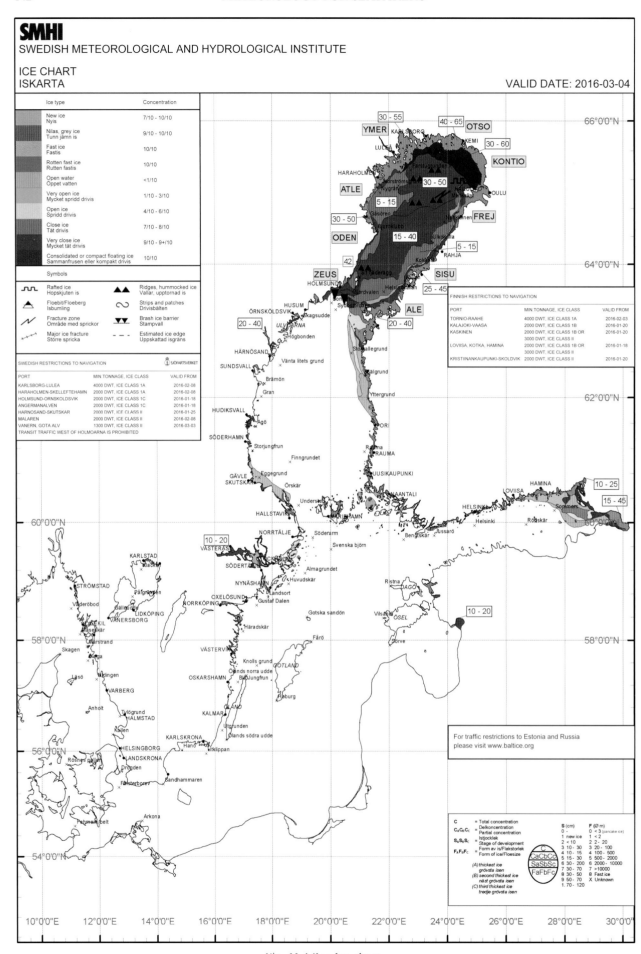

Fig. 11.4 Sea Ice chart.

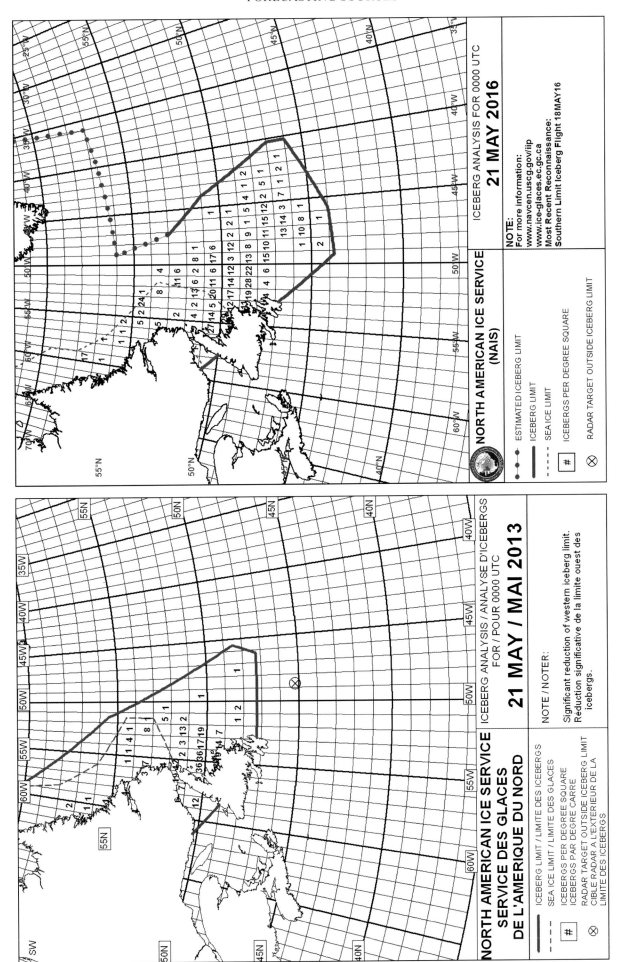

11.5 Iceberg charts.

UTILIZATION OF FACSIMILE DATA

Facsimile transmissions are a potential source of a large volume of data which can be incorporated into passage planning. In order to gain the benefits of the data, a careful study of the transmission schedules is necessary. However, the suitability or otherwise of the data may only become apparent when the actual charts have been received.

Surface analyses and prognostic charts received by the seafarer may be studied individually and then compared, thus establishing the state and the likely evolution of the atmosphere in terms of pressure systems, their movement, development and decay.

A review of Figs. 11.6, 11.7 and 11.8 illustrates the case. At 1200 UTC on March 25th 2016 (Fig 11.6), a frontal depression positioned at 42°N 167°E is forecast to move NNE in the next 24 hours. On the chart, details of the associated storm warning are noted and a fog warning, where warm air moves into higher latitudes over oceans with lower sea surface temperatures. A comparison of Fig. 11.6 and Fig 11.7, the 24 hour surface prognosis, indicates that at 1200 UTC on March 26th 2016 the centre of the frontal depression will be positioned at 47°N 170°E deepening to 952 hPa and it will have begun to occlude. On Fig 11.7, the plots of forecast surface wind speeds confirm the anticipated steeper horizontal pressure gradients as the system deepens. During the same time period, an anticyclone positioned at 35°N 165°E with central pressure value of 1028 hPa is forecast to move east, whereas to the east of Korea a slow moving anticyclone present on March 25th will be positioned over and to the south-east of Japan with multiple centres (Fig 11.7). Note the inclusion of additional isobars, the dashed lines, to assist in the interpretation of forecast systems.

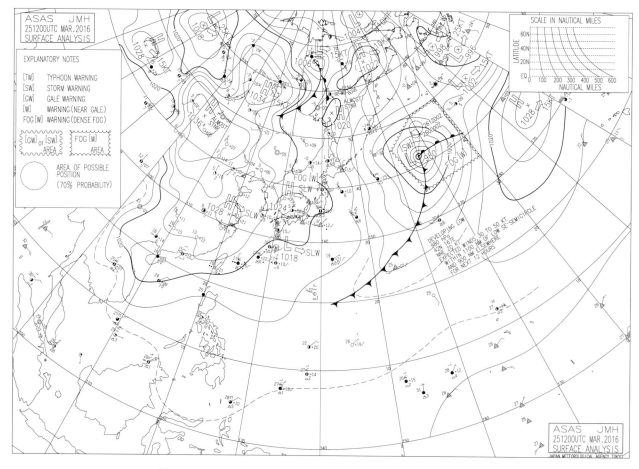

Fig. 11.6 Surface synoptic chart – March 25th 2016 1200 UTC.

The progress of the frontal depression in the following twenty-four hours can be traced by comparing the 24 (Fig 11.7) and 48 hour (Fig 11.8) surface prognostic charts. Thus at 1200 UTC on March 27th 2016, the centre of the frontal depression will be at 47°N 178°E with a central pressure value of 964 hPa as it will have begun to fill. It will continue to occlude, and the warm and cold fronts will have moved east simultaneously pivoting around the centre of the system. On Fig 11.8, in the western North Pacific the anticyclone which was positioned in and around Japan on March 26th will have moved to 31°N 161°E, its central pressure value increasing to 1026 hPa. Thirty knot winds are forecast on the north east-periphery of the anticyclone, where the horizontal pressure gradients are steeper in conjunction with the occluding frontal depression positioned to the north-east.

Further analysis of Figs. 11.6, 11 .7 and 11.8 would confirm areas of the North Pacific Ocean where the Trade Winds conditions are present and in the China Sea the North-East Monsoon.

The information extracted from the surface analyses and prognostic charts may be supplemented with information from other facsimile charts. For example the wave prognostic chart will define the anticipated wind and swell wave conditions over an ocean area (Fig. 10.17), and satellite images or nephanalyses will show the associated cloud forms as they develop.

As with weather bulletin data, the updating of all facsimile charts is important and relatively easy with the frequent transmission of data. Thus within twenty-four hours of receiving a 48 hour surface prognostic chart the seafarer will receive a 24 hour chart. The latter will be more reliable in terms of its representation of the future state of the atmosphere, a factor reflecting the limitations of the numerical models to forecast accurately for extended periods.

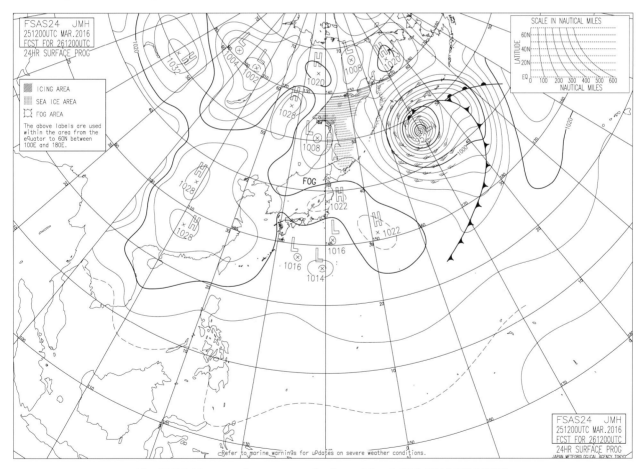

Fig. 11.7 24 hour surface prognostic chart – March 26th 2016 1200 UTC.

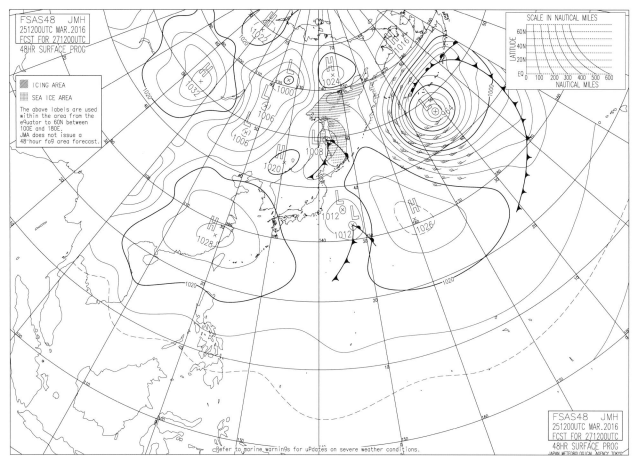

Fig. 11.8 48 hour surface prognostic chart – March 27th 2016 1200 UTC.

Facsimile charts and weather bulletins can be regarded as being complementary. The plain language forecast (Part 3) of the bulletin is a more detailed summary of forecast weather conditions for areas of the ocean, in terms of variations of wind speed and direction, precipitation and visibility, whilst the surface prognostic charts show the future position of pressure systems and fronts. Thus the details of the forecast in the bulletin amplify the data derived by comparing the surface analysis and 24 hour prognostic chart.

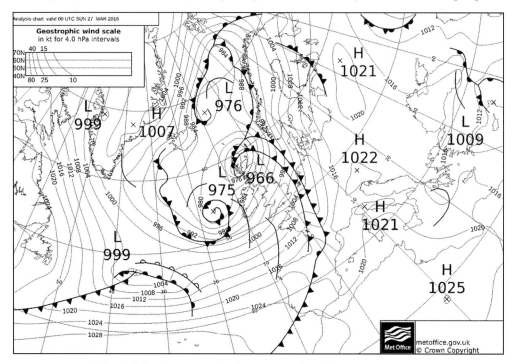

Fig. 11.9 Surface synoptic charts – March 27th 2016 0000 UTC

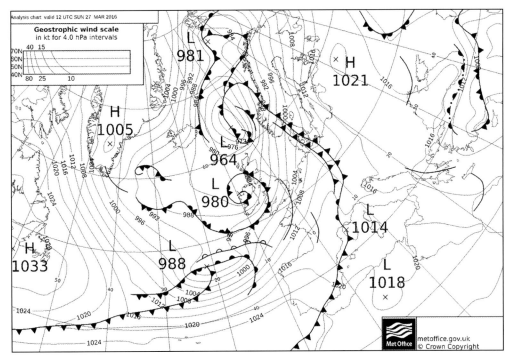

Fig. 11.10 Surface synoptic chart – March 27th 2016 1200 UTC

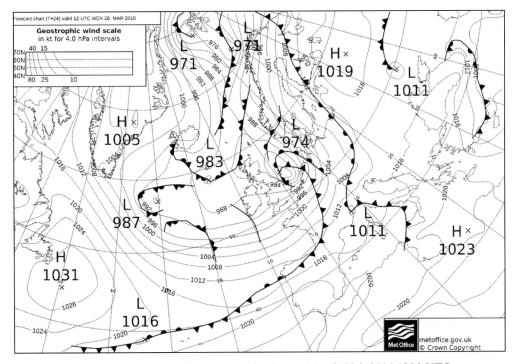

Fig. 11.11 24 hour surface prognostic chart – March 28th 2016 1200 UTC

As tabulated in the weather sequences for a frontal depression in Chapter 8, with the transit of the system initially the atmospheric pressure will decrease and later increase. By observing and recording the pressure, the progress of the system can be monitored, as illustrated by the readings recorded on March 28th 2016 by a lightship positioned in the English Channel (Fig. 11.12) and also the observations collected by a land-based observer in the United Kingdom (Fig 11.13). Observations collected can also be used to confirm the accuracy of the forecast with respect to the speed of movement of the system and wind conditions.

For example, the surface synoptic chart for March 27th 2016 0000 UTC (Fig. 11.9) shows a frontal depression positioned at 48°N 36°W with a central pressure value of 999 hPa. Correlating this information with the synoptic chart for March 27th 2016 1200 UTC (Fig. 11.10), the 24 hour surface prognostic chart valid for March 28th 2016 (Fig. 11.11) and information noted in the bulletins of Tables 11.2 and 11.3, indicates that the frontal depression is expected to deepen and transit the United Kingdom and adjacent coastal waters.

Finally it should be stressed that, notwithstanding the availability of both weather bulletin and facsimile data, the direct observation of the immediate environment should not be forgotten; it should be used to assess the accuracy of both types of data, and thus interpret and modify the forecast.

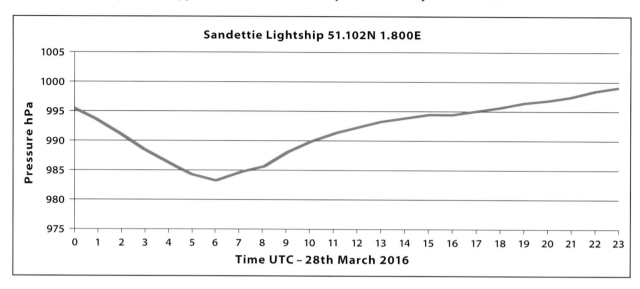

Fig. 11.12 Pressure readings – Sandettie Lightship March 28th 2016
Compiled from raw data supplied by the Met Office

© Crown Copyright, Met Office.
Courtesy of National Climatic Data Centre

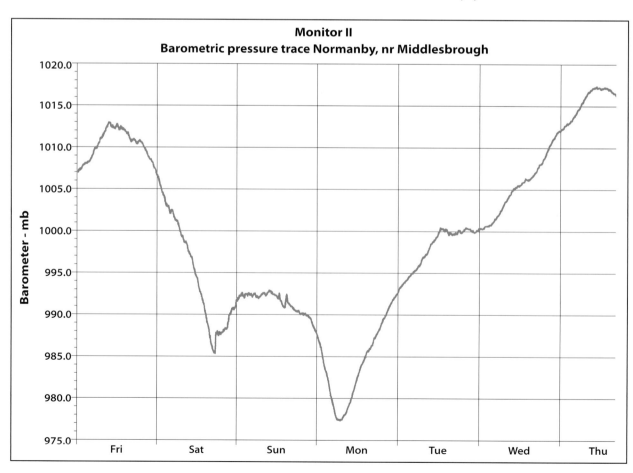

Fig. 11.13 Pressure readings – March 25th to March 31st 2016, Normanby, 53.4°N 0.5°W

CLIMATIC DATA

Routeing Charts

Routeing charts for the oceans are published by many organizations and Table 11.6 lists the data which is based on observations collected over a number of years. The charts present a summary of the climatic conditions for a particular month in a standard form (e.g. mean pressure distribution shown by isobars), and also use graphical and numerical forms of presentation. Some of the data may be shown on inset charts and tables.

Climatic conditions	Presentation
Air pressure, pressure systems	Isobars and, in some cases, frequencies of deep depressions with depiction of storm tracks.
Surface wind	Frequency distributions of wind direction in 8 points of the compass (wind rose) and wind speed, frequencies of gales and storms.
Air, sea surface, temperatures and dew-point temperatures	Isotherms of mean values.
Visibility	Frequencies of visibility.
Tropical cyclones	Tracks of individual cyclones and distribution for the months of the year.
Surface currents	Predominant direction and rate of current.
Sea ice and icebergs	The geographical distribution of sea ice and icebergs.

Table 11.6 Routeing chart data.

Sailing Directions

Sailing directions or pilot books also include climatic data on conditions at sea and climatic tables for observing stations in coastal areas. Information on mean pressure distribution for certain months, wind roses, winds of Force 7 or greater, tracks of depressions and tropical cyclones, reduced visibility, and fog are presented in chart form. The supporting descriptions summarise the charts and provide additional data on cloud, precipitation, thunderstorms, air temperatures, relative humidity, fronts and local winds. Ocean currents, sea and swell conditions, sea surface temperature, and ice are included.

Although the data from both routeing charts and sailing directions are climatic in nature, they are of value to the seafarer. However, intelligent interpretation of the data is needed, particularly in those areas where the weather conditions change frequently over relatively short periods.

If the seafarer makes full use of on board observations, weather bulletins, facsimile services and climatic data, then he will have the best information possible to ensure a safe passage.

APPENDIX 1

TYPHOON FAYE AND EXTREME STORM WAVES

Chapter 9 discussed tropical cyclones and suggested a guide for ships. The account below, reproduced by kind permission of the *US Mariners Weather Log* (vol. 27), highlights the advanced warning provided by the swell, which in this case was violently emphasized by an extreme storm wave, and the real danger and difficulties encountered at sea by vessels entering a tropical cyclone. The ship's barogram is included in Chapter 2, but other photographs of damage, which were included in the article, have been omitted.

SS MOBILE AND TYPHOON FAYE

The following is a summary of Captain Cordes' account of his voyage from Kaoshing, Taiwan, to the US Naval Base, Subic Bay, Philippine Islands in ss MOBILE. Times are LMT.

"The MOBILE departed Kaoshing at 0742 on 23 August, 1982, for Subic Bay. A full profile of containers was aboard with truck chassis in bundles of two lashed on top of the containers on hatches No. 1, No. 2, and No. 9. Wooden 4 × 4's were laid across the tops of the containers to distribute the load of the chassis.

The MOBILE proceeded along the west coast of Luzon, which reduced the general height of the sea from the easterly wind. The weather broadcast from Hong Kong located Typhoon Faye 13°N 118°E at 0500 on 24 August, 1982, and forecast it to remain essentially stationary for the next 24 hours. There was no evidence of the typhoon other than a slowly increasing southerly swell.

At 1500 on 24 August the ship was struck by an extreme storm wave. The ship pitched steeply down into the trough ahead of the wave, slammed into the wave, and pitched upward rapidly. When the wave passed under the stern, the MOBILE was again slammed into the sea. After a short period of violent pitching the ship began to ride normally again. This occurred at 15°09'N, 119°52'E. A container on hatch No. 1 burst open and some of the mail bags were washed over the side when the extreme wave came aboard. A contributing factor to the damage was the large moment arm of the heavy chassis on the top of the containers well forward of the ship's centre of movement. When the violent pitching and slamming occurred, the chassis became like huge hammers.

At about 1600 the US Military Sealift Command at Subic Bay was contacted for the latest weather forecast. Typhoon Faye's position was the same. It was expected to pass Subic 60–70 miles to the west at its closest point of approach at about 2300. Speed was reduced to allow better reaction to any other extreme waves that might be coming and the vessel proceeded on course to Subic, anticipating arrival about 1900, well ahead of the typhoon.

At 1630 the weather was good, light overcast of the usual monsoon type, the wind was east-northeast at 10–15 knots, and there was a moderate southerly swell. By 1700 the wind had increased to 50 knots and an ominous line of low dark clouds appeared on the horizon ahead. Typhoon Faye had not been stationary at all, but had continued to move northeastward at about 15–20 knots since the morning report.

It was fortunate that the wind was from the east because the MOBILE was only three miles west of Capones Island. A westerly course was tried to get the ship away from land in anticipation of the wind reversing direction on the opposite side of the typhoon eye.

At 1730, the wind was from the east at more than 60 knots and the barometer was dropping precipitously. The ship would not hold a westerly course in the wind and high sea and swell. The ship's head fell off to the southwest, and at 1805 maximum speed was needed to hold the southwesterly course. A fully-loaded container ship in a storm has all the attributes of a square-rigged sailing ship with all sails set.

Blinding rain reduced visibility to zero and made the radars ineffective and by 1830, the wind speed had increased to a steady 80–90 knots with gusts estimated at 100 knots. However, the ship was now riding well with *wind and sea on the stern. Speed cautiously reduced to the minimum for steerage. Both radar antennae refused to turn in the high wind.*

At about 1945 (1145Z) the MOBILE entered the eye of the typhoon. One radar began functioning and the eye could be clearly seen encircling the ship, but the heavy rain of the storm attenuated the radar so that targets beyond the eyewall could not be detected. The Loran C gave a reasonable fix (within five to six miles in this area). To avoid any chance of grounding, it was decided to maintain the southwesterly course. The position of the eye of the typhoon was transmitted to Hong Kong and they responded with typhoon advisory No. 12 relocating Typhoon Faye. At about the same time the ss PRESIDENT POLK, 20 miles to the northeast, reported 80-knot winds.

At about 2030 the MOBILE entered the southern wall of the eye. The wind jumped to an average steady wind of 80–90 knots with gusts well over 100 knots. Spray and torrential rain reduced visibility to near zero and seas were mountainous. The radar ceased functioning again as the wind increased. Many of the ship's parts and fittings could be seen blowing or bending in the wind with water coming across the ship. The masthead and range lights were lost around 2045. Finally, the storm began to decrease about 2130 with wind under 50 knots and, for the remainder of the night, the wind gradually decreased.

The following is a list of the damage:

(a) Six chassis lost overboard

(b) Four chassis damaged

(c) Four containers destroyed on No. 1 hatch

(d) Eight containers on No. 2 hatch damaged

(e) Both masthead and range lights destroyed

(f) HF radio antenna bent back 45 degrees

(g) All wire antennae destroyed

(h) Both cranes inoperative

(i) Anchor windlass controller sheared off at deck level

(j) Vent pipe sheared off allowing bosun's stores over No. 2 deeptank to flood with 3 ft of water."

This account was published again in *Seaways*, May 1984, from whom the permission has been given to publish the comments of Captain E. W. S. Gill:

"The remarkable thing regarding this incident is that even in this era of space-age technology, the position of tropical cyclones cannot be guaranteed. In this particular case the typhoon was surely within strategic radar range of the US Naval Base at Subic Bay, as well as any strategic radars at Manila. Yet the actual position and movement of the storm was undetected until the MOBILE reached the centre, and confirmed it.

It also emphasises the importance of an accurate predicted pressure inside the cyclone. This can be of inestimable value to a seaman in determining the position of one's ship, relative to the storm. Personally, I have always endeavoured to get below (equatorially) a tropical cyclone, rather than rely upon a steady predicted movement of the cyclone as indicated by meteorological reports. It's safer.

The comments in the article regarding the inability of even large and powerful container vessels to manoeuvre as one desires, once the storm has a ship in its grip, can be vouched for by myself. I had previously noted that container vessels with their large windage areas can be extremely difficult to keep on course that would normally be best for the ship under the hurricane force weather conditions.

With a quarterly wind and sea, the tendency for the ship to try to broach to is enormous. The large application of helm to prevent the ship turning has the counter affect of slowing the ship down and losing steerage way so that it is imperative that the steering be closely watched under these conditions.

With the same strength of wind and sea from ahead there is again the tendency for the ship to pay off the wind rather violently, and any increase of speed merely increases the pounding effect. The only thing to do under these circumstances, and providing there is sufficient sea room, is to cut and run only never leave it too late to make this decision, for the turning of a ship in tumultuous sea, combined with extreme wind heel factors could put the ship over to an impossible angle".

EXTREME STORM WAVES

Extreme Storm Waves (ESW), one of which struck ss MOBILE, arise from the combined effects of meteorological and oceanographic conditions (Plate 44). The following table, produced by the US National Weather Service, is a guide to the different types of waves.

Description	*Comment*
Steep, white water on crest; "hole-in-the-sea" in front; usually occurs as a single large wave aligned with other waves in the seaway.	Known to occur in strong, increasing winds. Tend to reoccur as "every 7th wave" among the larger waves in the seaway. These large, steep waves are apt to cause local damage to a ship or its lifeboats, accommodation ladders etc.
Group of large waves–usually three, with the second wave being highest. The group is aligned with other large waves in the seaway.	Likely to occur in strong winds which have peaked or begun to decrease in strength. During the period of peak winds these waves are apt to be much higher than the other large waves in the seaway. This wave group can produce severe hull girder bending stresses.
Group of large waves, usually three, misaligned by 30 degrees or more to local wind-driven seaway.	Wave group intrudes unexpectedly into local seaway and produces locally high wave crests at intersections with the largest waves in the local wind-driven seaway. The wave group tends to cause severe roll response by a ship.
Large breaking wave(s) intruding into local seaway at angles up to 50 degrees from direction of local wind driven seaway; of unusual height with respect to other waves in the local seaway.	Frequently characterized as a "rogue" wave(s). Very dangerous due to associated "wall of water", and the fact that they appear unexpectedly (Note 1).

Note 1: The "rogue" waves which occur in the Agulhas Current off the south-east coast of Africa have this characteristic, but these are not the waves being described here.

Table A1 Classification of extreme storm waves.

Alternative terms for exceptionally large waves include freak, extreme storm and rogue waves. The observation of such waves has been much improved by the use of sensors on offshore units, buoys and satellites. Extensive research has and is being conducted to establish factors which may control their development. With reference to the footnote of Table A1, development can be due to the interaction between opposing swell waves and a strong ocean current. Other theories include the interaction of waves in crossing seas, and the focussing of energy within a wave train.

Given the adverse conditions generated by these waves and the need to consider the criteria used in the design of structures to be deployed in the marine environment, research continues with the general aim of improving the forecasting of the occurrence of such waves. Readers may explore the issue by referring to research papers and books where, in the latter, the generation of tsunamis is often included.

WORLD METEOROLOGICAL ORGANIZATION
GLOBAL MARITIME DISTRESS AND SAFETY SYSTEM METAREAS

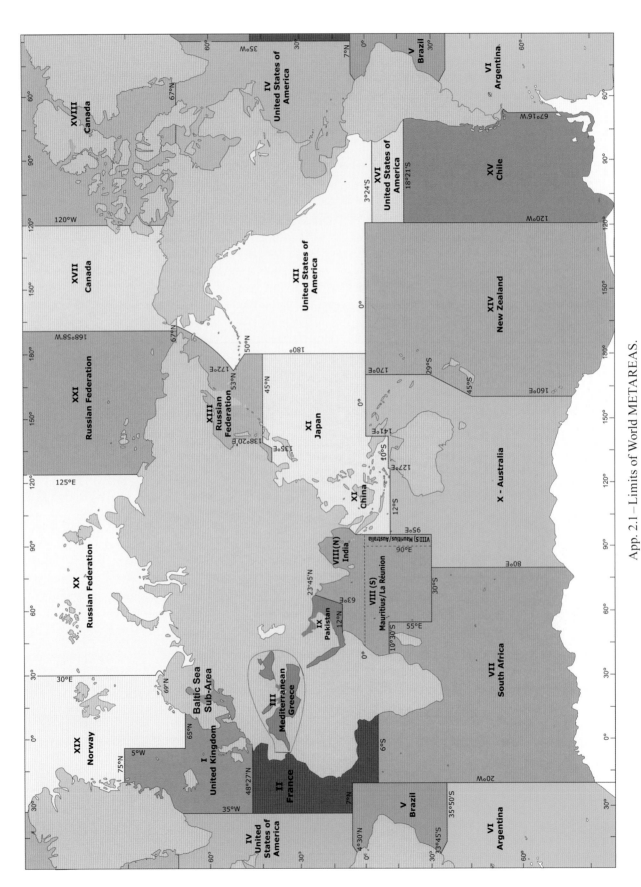

App. 2.1 – Limits of World METAREAS.

WORLD METEOROLOGICAL ORGANIZATION
GLOBAL MARITIME DISTRESS AND SAFETY SYSTEM METAREAS

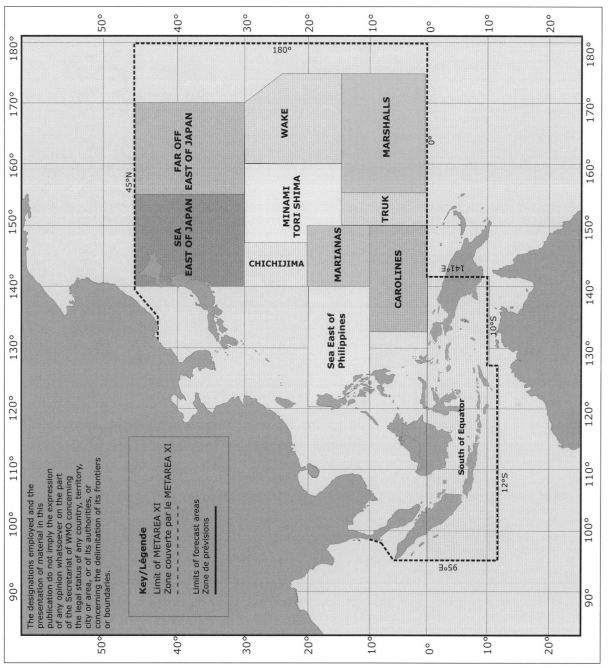

App. 2.2 – Limit of METAREA XI.

APPENDIX 3

OTHER SOURCES OF INFORMATION

Several sources of information have been referred to in this book. The following organizations and works of references are also suggested to those seafarers who wish to extend their understanding of the atmosphere and the oceans of the world.

ORGANIZATIONS

1. The UK Meteorological Office, Fitzroy Road, EXETER, Devon EX1 3PB. It has a library and meteorological archives. Seafarers seeking advice should contact Customer Services or visit its website.

    Telephone: +44 (0)370 900 0100

    e-mail: enquiries@metoffice.gov.uk

    website: www.metoffice.gov.uk

2. The Royal Meteorological Society, 104 Oxford Road, READING RG1 7LL, is a private organization whose membership includes professional meteorologists and anyone interested in meteorology. The Society has an excellent range of journals and a library. Any seafarer who would like further information should contact the Secretary.

    Telephone: +44 (0)118 956 8500

    e-mail: info@rmets.org

    website: www.rmets.org

BIBLIOGRAPHY

A selection of books likely to be of particular interest to seafarers is given below. Seafarers wishing to buy any of the books may contact The Marine Society and Sea Cadets, 202 Lambeth Road, LONDON SE1 7JW

    Telephone: +44 (0)207 654 7000

    website: www.marine-society.org

BOOKS

Authors and titles		Publishers
1. The Marine Observer's Guide	(Met 0 477A)	
2. Ship's Code and Decode Book	(Met 0 509)	
3. Marine Observer's Handbook	(Met 0 1016)	Her Majesty's Stationery
4. Meteorology for Mariners	(Met 0 895)	Office (HMSO)
5. The Mariner's Handbook	(NP 100)	
6. Ocean Passages of the World	(NP 136)	
Ahrens, C. D.	Meteorology Today (2012)	Cengage Learning
Ambaum, M. H. P.	Thermal Physics of the Atmosphere (2010)	Wiley-Blackwell
Barry, R. G. and Chorley, R. J.	Atmosphere, Weather and Climate (2009)	Routledge
Chang, C. P. and Krishnamurti, T. N.	Monsoon Meteorology (1988)	Oxford University Press
Emanuel, K.	Divine Wind: The History and Science of Hurricanes (2005)	Oxford University Press
Galvin, J. F. P.	An Introduction to the Meteorology and Climate of the Tropics (2015)	Wiley-Blackwell
McIlveen, J. F. Robin	Fundamentals of Weather and Climate (2010)	Cambridge University Press
Pelinovsky E. and Kharif C.	Extreme Ocean Waves (2017)	Springer
Talley, L., Pickard, G. L., Emery, W. J. and Swift, J. H.	Descriptive Physical Oceanography (2011)	Academic Press
Sturman, A. and Tapper, N.	The Weather and Climate of Australia and New Zealand (2005)	Oxford University Press
Wallace, J. M. and Hobbs, P. V.	Atmospheric Science: An Introductory Survey (2006)	Academic Press

INDEX